Science,
Ethics,
and
Food

Science, Ethics, and Food

Papers and Proceedings

of a Colloquium

Organized by the

Smithsonian Institution

Edited by Brian W.J. LeMay

Smithsonian Institution Press
Washington and London

Copyright © 1988 by Smithsonian
Institution
All rights reserved

Lisa Buck Vann, *Designer*

Library of Congress Cataloging-in-
Publication data

Science, ethics, and food.

Includes index.
1. Food supply—Congresses.
2. Food industry and trade—Moral and
ethical aspects—Congresses.
3. Agricultural innovations—Congresses.
I. Smithsonian Institution.
II. LeMay, Brian W. J.
III. Series.
HD9000.5.S377   1988   363.8   88-18399
ISBN 0-87474-605-1 (pbk.)

All photographs are courtesy of the
Smithsonian Institution.

Manufactured in the United States of
America

∞ The paper used in this publication meets
the minimum requirements of the American
National Standard for Permanence of Paper
for Printed Library Materials Z39.48-1984.

Contents

7   Foreword
        Robert McC. Adams, Secretary, Smithsonian Institution

11   Acknowledgments

        SESSION I
14   On the Protection and Promotion of the Right to Food:
        An Ethical Reflection
        William J. Byron, S.J., President, The Catholic University of America

31   Summary of Discussions
        DISCUSSANTS
        C. Gopalan, Director General, The Nutrition Foundation of India
        Robert W. Kates, Director, World Hunger Program, Brown University
        J. E. Dutra DeOliveira, Faculty of Medicine, Sao Paolo, Brazil

        SESSION II
38   The Innovative Environment for Increased Food
        Production in Africa
        Thomas R. Odhiambo, Director, The International Centre of Insect
        Physiology and Ecology

52 Summary of Discussions
    DISCUSSANTS
    Nyle C. Brady, Senior Assistant Administrator, Bureau for Science and
        Technology, U.S. Agency for International Development
    Ruben L. Villareal, Dean, College of Agriculture, University of the
        Philippines at Los Baños
    Mamdouth Gabr, President, International Union of Nutrition Sciences

    SESSION III
58 Food Entitlements and Economic Chains
    Amartya K. Sen, Professor of Economics and Philosophy, Harvard University
71 Summary of Discussions
    DISCUSSANTS
    Ignacio Narvaez Morales, Program Director, Global 2000 Inc.
    Shlomo Reutlinger, Senior Economist, The World Bank
    Mogens Jul, Associate Professor Emeritus, Department of Food Preservation,
        The Royal Veterinary and Agricultural University of Denmark

    SESSION IV
78 Toward an Ethical Redistribution of Food and
        Agricultural Science
    John W. Mellor, Director, International Food Policy Research Institute
93 Summary of Discussions
    DISCUSSANTS
    Qing Wang, Director, Food Research Institute
    Setijati Sastrapradja, Director and Botanist, Centre for Research in Biotechnology

    SESSION V
98 Summary of Discussions
    Led by M. S. Swaminathan

    APPENDIX A
120 Our Common Agricultural Future
    Acceptance Address on the Occasion of the Presentation of the General
        Foods World Food Prize
    M. S. Swaminathan

    APPENDIX B
131 Notes on Contributors

    APPENDIX C
135 List of Colloquium Participants

Foreword

Robert McC. Adams
Secretary
SMITHSONIAN
INSTITUTION

About a year before the General Foods World Food Prize was first scheduled to be awarded, we were approached by the General Foods Fund, who inquired whether we would be interested in organizing a scholarly colloquium on global food issues, to take place at the time of the presentation of the award. It seemed a timely opportunity for us to solicit the wisdom of colleagues from a number of different fields and try to address in some modest way the heartbreaking problems that stare at us from the pages of our daily newspapers. We ultimately arranged not only to organize the colloquium devoted to "Science, Ethics, and Food," which is documented in this volume, but also to have the World Food Prize itself presented here at the Smithsonian.

We succeeded in gathering together for the colloquium a remarkably diverse and distinguished group of government leaders, policymakers, food scientists, scholars, and food industry representatives from around the world. Quite apart from the papers presented at the colloquium and the perceptive comments they elicited during formal discussion periods, the gathering provided an opportunity for this formidable group to interact, to establish new or renewed contacts, and to exchange ideas informally—over coffee in the back room of our Hirshhorn Museum, during luncheon in our new International Center, or while sampling foods from around the world at a reception following the award ceremony in our Natural History Museum.

As our speakers, we were fortunate to be able to hear from four outstanding authorities in their fields: Father William J. Byron, President of The Catholic University of America; Thomas R. Odhiambo, Director of the International Centre of Insect Physiology and Ecology in Nairobi; Amartya K. Sen, Professor of Economics and Philosophy at Harvard University; and John W. Mellor, Director of the International Food Policy Research Institute. Serving as moderators for the first day's discussions were Jessica Tuchman Mathews, Vice President and Research Director of the World Resources Institute, and Robert Paarlberg, Associate Professor of Political Science at Wellesley College.

On the day following the presentation of colloquium papers and the World Food Prize, Dr. M. S. Swaminathan chaired a session designed to draw conclusions and prepare some recommendations for future steps that might be taken to address critical food problems facing the world today. It was encouraging to note the earnestness, specificity, and flexibility of approach emerging in these discussions with ever greater clarity; however, the magnitude of the challenges that appeared before us was indeed sobering.

In spite of successes like those celebrated by the World Food Prize, recurrent hunger and unremitting malnutrition remain virtually as widespread and dangerous as ever. In Dr. Odhiambo's colloquium paper, for example, we are quietly warned of the tragic trends in African food production that have taken place over essentially the same period that has seen triumphs over famine in southern and southeastern Asia. Noting Africa's institutional deficiencies as well as shortages of trained scientists and specialists, Odhiambo draws our attention to the continent's endemic, devastating illnesses that have prevented agricultural exploitation of large areas of land there. Without a revolution in rural health care, he tells us, no long-term, sustainable system of agricultural production will be possible. At the same time, the problems of African agriculture cannot be solved by any quick technological fix. Though the Green Revolution has raised hopes about the power of technology to solve our agricultural problems, what is needed in this case are low-input, low-cost solutions that incorporate, rather than abandon, local subsistence practices and resources. Exporting technology alone, in other words, will not provide an answer. Western scientists and agronomists have a great deal to learn before they will have very much to teach.

There are reasons to be wary of the proposition that we in the West, troubled mainly by overproduction, have our agricultural act together, and have real help to offer in meeting the African challenge. In the United States, for example, a number of disturbing trends have accompanied the increasing dominance of agribusiness. By 1981, there were twenty-five thousand superfarms (with annual sales of $500,000 or more) that accounted for almost two-thirds of total net farm income. Farming itself has become a relatively minor, as well as subordinate, part of the agricultural system. In little more than a generation, the farmer's share of the retail price of a loaf of bread has dropped by more than half—to under 8 percent. The new, privately financed biotechnologies shorten the time required for breeding programs that formerly were largely publicly financed, further encouraging factorylike production and increasing the prospect of the linked marketing of agricultural chemicals and compatible seeds.

While positive in many respects, these developments direct us away from Africa's—and to some extent the world's—emergent needs. Agricultural experiment stations are induced to abandon varietal breeding and to concentrate on the handful of crops that dominate the world markets. Research on "minor" crops dwindles, in spite of its relevance not only for Africa, but for retaining the long-term protection of biodiversity.

As operators of smaller "family farms" continue to be driven out of agriculture and into the labor market, whence will come people with either the teaching or learning skills that the next great phase of the Green Revolution will surely require?

Unfortunately, troubling sitations like this, whether in the West, in Africa, or in Asia, often do not receive the attention they merit until circumstances approach crisis proportions. It was, of course, partly in order to focus the attention of our policymakers and the public on problems such as this that the World Food Prize was established. Problems this broad cannot be properly addressed, however, until they are clearly articulated and formulated in terms that draw upon an equally broad range of disciplines, perspectives, and arenas of action. Because of the international, multidisciplinary scope of Smithsonian activities and collections, we have long taken a more than theoretical interest in issues such as these, and it was therefore appropriate that the Smithsonian should provide a forum for the wide-ranging examination of food issues in the "Science, Ethics, and Food" colloquium. For their support of the colloquium and for the publication of this volume, we would like to express our appreciation to the General Foods Fund, Inc.; the Winrock International Institute for Agricultural Development; and the International Rice Research Institute.

Acknowledgments

Directors for the Smithsonian Institution colloquium documented in this volume were Carla M. Borden, associate director of the Office of Interdisciplinary Studies, and Brian W. J. LeMay, assistant director, Office of International Relations. The organizers of this colloquium particularly wish to acknowledge the invaluable contributions of Edward L. Williams, administrator of the General Foods World Food Prize at Winrock International. Mr. Williams displayed a willingness to share equally in all organizational duties of the colloquium and provided a plenitude of specialized knowledge that proved essential to the mounting of this event. Wilton Dillon, director of the Office of Interdisciplinary Studies, was instrumental in bringing about the Smithsonian's involvement in this project and was responsible for formulating the basic theme and title of the colloquium. Initial research for the colloquium was conducted by Neil G. Kotler, special assistant to the director, Office of Interdisciplinary Studies. In developing the program for the colloquium, we were indebted to the four featured contributors, to M. S. Swaminathan, 1987 General Foods World Food Prize laureate, and to our two moderators, whose wise counsel and sensible advice we sought on a number of occasions. Shouldering major burdens of planning and execution of the colloquium were Paul F. Hopper, corporate director of scientific affairs at General Foods; A. S. Clausi, chairman of the General Foods World Food Prize Council of Advisors; their perspicacious assistants at General Foods, Peggy O'Neill and Pamela Holdstock; Anne Swartzel, charming, patient, and efficient assistant to Edward L. Williams at Winrock International; and the remarkable Stephany F. Knight, our special projects manager, in whose capable hands we confidently left many of the arrangements for colloquium and World Food Prize events at the Smithsonian. David Challinor, then-assistant secretary for research and acting director of the Directorate of International Activities at the Smithsonian, should also be acknowledged as the voice of pure reason throughout our planning and negotiations. Thanks should be given also to our Gallic intern, Sigisbert Ratier.

The remarks included in the Foreword by Robert McC. Adams appeared in somewhat different forms in his column "Smithsonian Horizons" in the February, 1988 issue of *Smithsonian* magazine, and in a recent issue of the *International Agriculture Update*, published by the Office of International Agriculture, University of Illinois at Urbana-Champaign.

Father Byron's paper is to appear in the Fall 1988 issue of *Social Thought*, published by Catholic Charities USA.

Dr. Mellor's paper was published in the 1987 *Report* for the International Food Policy Research Institute.

SESSION I

On the Protection and Promotion of the Right to Food

An Ethical Reflection

Being human is both a right and an achievement. A human being, by virtue of his or her being human, is possessed of both rights and the potential to achieve. The rights are an innate part of being human. They are known as human rights. The potential to achieve points to the presence of responsibilities, also resident in the being human. Each individual has responsibilities for personal human development.

Failure to realize one's human potential—a failure to achieve—does not constitute an abdication or loss of human rights. These remain resident in the person in function of that person's being human; they reflect human dignity.

Human rights are quite basic. They are present whenever human beings are. They do not depend on what human beings do; they are there precisely because human beings are human. Human rights are therefore present to the material side of human existence. Each person, in function of being human, has a right to those material necessities without which human life and human dignity cannot be sustained.

There are rights, of course, in orders higher than that of material survival. But rights at higher levels are meaningless without the acknowledged presence and effective protection of rights at the material level of existence consistent with basic human dignity. One such right is the right to food.

Just as hunger can be viewed as an intensely personal or individual problem, so too can it be seen as a wider, even worldwide, societal problem. Both perspectives are necessary if the rights and responsibilities associated with the problem are to be held in proper context. Analysis of

William J. Byron, S.J.
President
THE CATHOLIC
UNIVERSITY OF
AMERICA

an individual's hunger and right to food based on that individual's possession of human dignity can lead to an unbalanced, individualistic approach to alleviation of the problem. Shifting the ground of ethical reflection from individual human rights to a broader concept of the common good may introduce a welcome communitarian dimension to the analysis, but it may also open the door to the conclusion that an ethic built on rights (not the common good) is more cause than cure of a problem so complicated as that of hunger in the world. The unfettered, individualistic exercise of human rights and the unregulated play of free markets are in no small way responsible for the problem of world hunger, according to what might be called the communitarian view.

I propose to offer in this paper a coherent but not comprehensive ethical reflection that goes well beyond classic liberalism's emphasis on individual rights and locates the human person, possessed of a right to food, in a broader communitarian context. I regard life in community as indispensable for the development of the human person. The communitarian context is essential for the realization of individual human potential and the protection of individual human rights.

The Principles

The basic principle underlying the right to food is the principle of human dignity. In saying this I would want to identify the principle of human dignity as the bedrock of a body of doctrine that has come to be known as Catholic Social Teaching. It is not, however, specifically or properly Catholic in any exclusionary, confessional sense. It is a universal principle available to human reason, but illumined, in the Catholic view, by revelation.

The principle of human dignity is referenced, of course, in many secular sources. Take, for example, the Charter of the United Nations, which affirms a "faith in fundamental human rights, in the dignity and worth of the human person, in the equal rights of men and women and of nations large and small."

Human dignity is the natural endowment of every human person. All human rights relate to and depend upon it. Hence human dignity is a principle from which all human rights, including the right to food, flow. But human dignity does not exist in some disembodied, abstract, splendid isolation. It requires association with other persons. Such association is essential for human development. Human persons are quite evidently social beings made for the company of others, capable of communicating, cooperating, procreating with others, capable of love and care for others. The realization of these capabilities requires association with others. This principle of association flows from the principle of human dignity. So does the principle of participation in community, participation as an associate of equal dignity with other human beings. To deny participation within the community to individuals or groups who have a right to be there is to disrespect, disregard, or even attack directly their human dignity. Every person, by virtue of being a person, has a right not to be marginalized, shut out, put down, isolated. Without participation, human development does not happen.

When the Catholic bishops of the United States rearticulated these principles of human dignity, association, and participation in their recent document *Economic Justice for All: Pastoral Letter on Catholic Social Teaching and the U.S. Economy*,[1] they added two other principles that bear on the topic at hand. These are the principles of subsidiarity and the principle of preference for the poor.

"Subsidiarity" might best be explained by simply citing the classic expression of this principle as stated in the encyclical *Quadragesimo Anno* of 1931. The bishops quote the passage in their economics pastoral:

> Just as it is gravely wrong to take from individuals what they can accomplish by their own initiative and industry and give it to the community, so also it is an injustice and at the same time a grave evil and disturbance of right order to assign to a greater and higher association what lesser and subordinate organizations can do. For every social activity ought of its very nature to furnish help [*subsidium*] to the members of the body social, and never destroy and absorb them.[2]

This principle will protect freedom, initiative, and creativity in the community. It will also justify subsidies where they are really needed.

The principle of preference for the poor is a biblically based, specifically Christian norm that measures the conformity of the values and choices of the Christian to the values and choices of Christ. Reasoned argument, apart from revelation, would produce the same conclusion. When it comes to protection of human dignity, preference should be directed to the point of greatest vulnerability, to situations of exclusion from association and denial of participation. In terms of economic survival, the poor are most vulnerable. Hence the poor need preferential protection. In terms of hunger—the most urgent form of poverty—the poor who are starving or chronically malnourished deserve preferential protection. Human dignity requires it. Starvation and chronic undernutrition are assaults on human dignity.

Protection of the Right to Food

The meaning I attach to the phrase "right to food" is that used in the House-Senate Concurrent Resolution Declaring as National Policy the Right to Food.

> Resolved that:
> Every person in this country and throughout the world has the right to food—the right to a nutritionally adequate diet—that this right is henceforth to be recognized as a cornerstone of U.S. policy.[3]

In Congressional Hearings on the Right-to-Food Resolution, Dr. Robert M. Cavanaugh, President of International Food Technology, Inc., of Greenville, Delaware, stated:

> My primary purpose in asking to be heard today is to make as emphatically as possible the point that passage of the first sentence of House Concurrent Resolution 393 would greatly facilitate essential discourse between nutritionists and economists, and later among various decisionmakers, because it contains language that marks a crucial paradigmatic shift.
>
> The focus becomes abruptly sharpened to a "nutritionally adequate diet," which has useful meaning, whereas the word "food" has almost none for vitally important planning purposes. Providing x million tons of "food" to allay the "hunger" of 464 million people is like trying to provide y million pounds of "medicine" to solve the "illness" of that many people.[4]

The meaning I attach to the "right to food" is contained in the key phrase—"the right to a nutritionally adequate diet." This is the claim any human person can make on the human community. This is not to say that this right imposes any obligation on someone else to produce the food, or to hand over food to anyone who might be inclined to assert the claim. Recall the opening sentence of this present paper: "Being human is both a right and an achievement." The right to achieve, I noted, points to responsibilities in the same human person possessed of human dignity and all derivative human rights. One of those responsibilities is to engage oneself with one's external, material environment, as well as to develop oneself intellectually and spiritually—that is, to cultivate both body and mind, and to interact with persons, ideas, and the material creation external to oneself in order to produce goods and services needed for the survival and development of self and the community. This is an elaborate description of employment, which is also a requirement of human dignity. The employment of some produces food for all. Those employed not in the production of food, but in producing other goods and services that the community needs or wants, derive income for their own use in the purchase of food and other necessities and wants.

The right to food does not, however, depend on employment performance. Failure to realize one's human potential for income and employment does not disqualify one from the human community. Failure to produce food or to earn sufficient income to buy food will lead to the form of deprivation known as hunger, but not to the loss of title to human dignity, to association, or to participation in community. Individuals or groups may find themselves frustrated, for whatever reason, in their efforts to achieve a level of being consistent with human dignity. When such failure leaves them without a "nutritionally adequate diet," they can assert a right to food. The community has the obligation to respond with emergency aid and other appropriate subsidies. The community has the further obligation to organize its systems and structures in ways that will enable persons to help themselves in the production of food or the generation of income sufficient to purchase food. Such arrangements will quite obviously protect the right to food by protecting the capacity to produce food or to earn income that can be applied to the purchase of food. Designing, constructing, and preserving these arrangements are community responsibilities.

Protection of the right to food extends quite logically to the protection of natural resources, of land, water, plant, and animal species. Conservation of food-producing resources, and research directed to the enhancement of the food-producing potential of those resources, fall within the scope of the broad community responsibility to protect the right to food. Not to be ignored in all of this is the question of distribution—another area of community responsibility that, if left unattended, can frustrate success on the production side.

The importance of research related to world hunger cannot be overestimated. The important distinction between nutrition and food underlies the challenge to the researcher to coax more nutrition out of less food, as well as to multiply crop yields through genetic manipulations that can also affect seeds and livestock embryos. The "field" for research that will be protective of the right to food extends well beyond agriculture to the unrealized potential of aquaculture. Society has the obligation to encourage the research talent needed for this task—to develop it, reward it, and recognize its work as contributory to the satisfaction of the societal obligation to protect the right to food.

In addition to the care and development of its food-producing systems, and in addition to its efforts to advance and apply nutrition science, a society intent on protecting the right to food will also have to attend to questions of ecological balance and trade equity. These issues are scientifically challenging, economically intricate, and geopolitically complex. Ethically, they are rooted in the principle of human dignity, related to the principles of association, participation, and subsidiarity, and strategically linked to a preferential protection for the hungry poor of the world. In an ethically sensitive world community, the realization that every human person possesses the right to a nutritionally adequate diet can fasten the attention of scientists, economists, and politicians on the problem of hunger in the human community.

Promotion of the Right to Food

I once edited a book entitled *The Causes of World Hunger*.[5] I undertook the project with an eye toward promoting the right to food. In my view, we all tend readily and regularly in our society to substitute blame for analysis. Extended causal analysis will move through considerations of geography and climate, resource abuse, population, poverty, politics, trade barriers, the colonial legacy, the unequal distribution of wealth and income in the world, the complexity and consequent neglect of agricultural development in many parts of the world, and the failure to establish a workable international system of grain reserves. The analysis will recognize the dead hand of tradition as contributing to poverty and hunger. It will inevitably note the absence of political will—in the developed and less developed nations—to deal effectively with the problem. Other causal considerations will surface as the analysis continues, but the single question of political will remains to be answered before an effective remedy to the scourge of hunger can be said to be at hand.

Political will is best organized around an issue. The issue must be articulated and communicated (the task of leadership) if a widely grounded (and therefore sufficiently strong) political will is to be established.

Articulation of the hunger issue in terms of a human right—the right to food understood as a claim to a nutritionally adequate diet—is an indispensable first step. Agreement on the existence of such a right wherever human life exists cannot be assumed. Without widespread acknowledgment and acceptance of that right, a solution to the problem of hunger is unthinkable. Hence the importance of the communication of a clearly articulated understanding of the right. Such communication is part of the business of promoting the right to food.

History is replete with instances of rights violated, indignities endured, and injustices overcome. In virtually every case, identification of the injustice had to be accompanied by a clear articulation of the right being violated and a persuasive communication of an understanding of that right as a basis of the exercise of remedial political will.

The hunger issue, understood as the violated right to a nutritionally adequate diet, is not a question of charity—an invitation to extend, for charitable and humanitarian motives, a helping hand. The hunger issue is a matter of justice. No one member of the community is exempt from the demands of justice (all members, in view of their shared human nature and common human dignity, are associated one with all others in justice relationships). But no one member can singly satisfy the requirements of justice relative to the question of world hunger. An individual respect for the right to food must enlarge itself to a communitarian concern for the protection of that right, as preamble to a community response to the problems produced by neglect or violations of that right.

Communities organize themselves into governments for the ordering and management of community affairs that cannot be handled effectively by individual persons, by lower levels of organization, or by voluntary charitable arrangements. Widespread hunger in the community is certainly a matter to be addressed by government—not by a total leave-it-to-government strategy, but by government participation in a cooperative response to violations of the right to food. But government will not respond without the impetus of political will resident in the citizenry. Moreover, government's response will require various forms of expenditure of revenues received from citizens. Political will, in the matter of hunger, will therefore involve both political choice (spend for hunger reduction instead of something else) and citizen assent to the payment of taxes (possibly higher taxes if other public expenditures cannot be reduced) for the alleviation of hunger in the community.

Organized as we are into nation states, rather than a unified world government, we cannot solve the hunger problem without worldwide international cooperation. Theoretically this seems possible. Human nature transcends national boundaries. So does human dignity. Human rights may not be evenly recognized and protected in all nation states, but they are equally resident in all human beings no matter where they are. Was Teilhard de Chardin wise or wishful, or both, in saying, "The Age of Nations is past. It remains for us now, if we do not wish to perish, to set aside the ancient prejudices and build the earth"? Our prejudices against the hungry must be broken. We—all of us in all nations—will have to pay for those prejudices one way or another. In fact, we will have to pay in a variety of ways, all peaceful, if we are to avoid paying in political unrest and violent uprisings in places where hunger is urgent and widespread. Those peaceful ways are a combination of taxes and transfers, voluntary contributions of time and money, citizen advocacy, career choice and vocational commitment aimed at the reduction of hunger in the world through research, increased food production, and improved food distribution. But this will not happen without general acknowledgment and widespread acceptance in the human community of the existence of the right to food.

The absence of a strong leadership voice and leadership insistence on the existence of a universal human right to food is regrettable. The pastoral leadership of Pope John Paul II is notable in affirming the right to food, but his voice alone cannot build the necessary political will. I was moved personally by imagery used by this pope in affirming the right to food in an address made in Mexico, in the rural region of Oaxaca, on 29 January 1979. In the presence of peasants, he reminded the rich and powerful classes that bread needed for the nourishment of poor families "lay hidden" in fields kept unproductive by decisions of large landowners. This situation, he said, "is not just, it is not human."[6] Other leadership voices in the religious, political, scientific, educational, and industrial spheres of influence must be heard if the right to food is to be recognized, protected, and promoted.

People who are troubled intellectually by any suggestion of a preference for the poor might reflect on evidence of human behavior in an altogether different area of life that transcends national and cultural differences and is virtually universal. Imagine a parent flanked by two children, a three year old and an adolescent, walking toward a busy intersection in New York, New Delhi, or any other place where vehicular traffic moves along in close proximity to pedestrians. The three year old breaks away from the parent's hand and darts into the cart path or street traffic. Without a thought, for the moment, of the older child's safety, the parent will move quickly—and preferentially—to protect the toddler. So will total strangers who see the problem. Why? Because of the evident vulnerability of the helpless child. Such a response is appropriate, correct, charitable, and just human behavior.

When the vulnerability of the hungry poor becomes sufficiently evident, the appropriate preferential protection will be more readily forthcoming. Effective promotion (by means of articulation, communication, depiction, and representation) of the right to food will build political will and encourage the consequent protection of the right to food. There is a limited role for government in this response. Political will should be strong enough to bring the response of government right up to its appropriate limit. But since political will resides in the same people who live in families, belong to churches and synagogues, and populate countless private sector organizations and voluntary associations, it might be presumed that their concern to do something about hunger through political means will carry over to touch the agenda of their nongovernmental activities. The encouragement of such a widespread response is the function of leadership in the promotion of the right to food. The eventual outcome will be the protection of the right to food.

Where There Is a Political Will, Will There Also Be a Political Way?

Multiple means will have to be employed if the human community is to move toward the goal of eliminating hunger. Economic, scientific, and technological means are crucially important. But the best technical means will remain untested and unapplied to the goal of ending hunger unless effective political means are in place and at work.

The Senate-House Concurrent Resolution on the Right to Food made mention of U.S. Secretary of State Henry Kissinger's declaration of a "bold objective" at the 1974 World Food Conference in Rome, namely, "that within a decade no child will go to bed hungry, that no family will fear for its next day's bread, and that no human being's future and capacities will be stunted by malnutrition." By way of preamble to the assertion of this objective, Secretary Kissinger noted: "The profound promise of our era is that for the first time we may have the technical capacity to free mankind from the scourge of hunger."[7] With the technical capacity already in place, it remains for us to design and apply the necessary political devices. At the World Food Conference in 1974, Secretary Kissinger pledged his government's willingness to "work cooperatively" with other nations toward the achievement of what is certainly a "bold objective." The nations represented at the World Food Conference should resolve, Kissinger urged, to "confront the challenge, not each other . . . and let us make global cooperation in food a model for our response to other challengers of an interdependent world—energy, inflation, population, protection of the environment."[8] No one will dispute the desirability of these goals. Nor should anyone dismiss as wishful thinking the possibility of learning, through cooperation in food, how to meet these other major challenges.

We have made little progress since 1974 in achieving global cooperation in food. The decade ending in 1984 saw a sharp rise in worldwide awareness of hunger, chiefly as a result of televised news reporting of famine in Africa. For the most part, however, the response to the problem took the form of emergency food aid. Political will was heightened; in the United States, it was translated into action by several political means and by private voluntary activity. But the volume of politically enacted relief assistance was not notable relative to the size of the American GNP and the enormity of the need overseas. And the private charitable response has been relatively short-lived, fading as graphic representations of the problem disappeared from the print and electronic news media.

Political means to the political goal of ending hunger are in need of design, redesign, and persistent application. What forms might they take?

In the U.S., Public Law 480 (Food for Peace) was enacted in 1954 in a domestic political environment characterized by huge domestic farm surpluses that could not be sold in this country. There was both a political and a commercial need to move grain surpluses to overseas markets. P.L. 480 food aid has been used to support U.S. foreign policy and national security goals. There are two basic ways in which P.L. 480 assistance is provided: (1.) long-term credit at attractive rates to needy nations to purchase U.S. farm products, and (2.) direct food donations. In recent years, the dollar value of the credit we extend to poor nations under P.L. 480 has far exceeded the dollar value of food donations. P.L. 480 is the basic policy tool we have for moving U.S. food into empty stomachs of poor people overseas.

The challenge of hunger will not be met if political strategists focus on P.L. 480 assistance and ignore the need to increase food production in the food-deficit nations. The emphasis in our foreign assistance policy has shifted since 1954 from food aid to combat hunger (while disposing of domestic surpluses and promoting foreign policy objectives) to a more recent concern with agriculture, as opposed to industrial development. In the early days of the U.S. Agency for International Development, industrial development was stressed. The idea was to draw surplus rural labor in poor countries into industrial jobs (usually located in cities); industrial workers would presumably purchase the product of the agricultural sector. By the late 1960s, it became clear that agricultural development was needed more urgently than industrial development. It also became clear, in due time, that small-scale agricultural development was wise in less developed nations with large numbers of poor people in rural areas. To observe that land reform is indicated if the rural poor are to participate in agricultural development is simply to note the nature of one key dimension of the political challenge.

In their valuable book, *To Feed This World*,[9] Sterling Wortman and Ralph W. Cummings, Jr., report Clifton Wharton's observation that politicians are professionals whose perspectives differ often from those of agricultural development professionals. Wharton sees "two political requirements (among others) for achieving significant sustained agricultural development: First, that the political leadership have a genuine *commitment* to the goal of agricultural development; and second, that they have an understanding of the process."[10] Both requisites for political leadership in this matter apply to politicians in rich nations and poor; they must cooperate in the design and application of anti-hunger strategies. This cooperation is more likely to occur if research results of scientific investigations are reduced to language the politicians can understand. Perhaps journalists have a role to play in interpreting the scientists to the policymakers. Political leaders cannot be expected to support research if they do not understand it. Nor will they attach a high priority to agricultural development if they do not see that it makes sense politically as well as economically.

A study paper prepared for consideration by participants in seminars on World Food Day, 1985, noted:

> This change in philosophy [from an emphasis on industrial to a stress on agricultural development] was reflected in the policies of such major aid agencies as the World Bank and the U.S. Agency for International Development (AID). Since the early 1970s, these agencies have focused on small-scale farmers seeking, through broad agricultural development programs, to provide resources such as better markets, price incentives, improvements in transportation, credit and land tenure arrangements, education, and agricultural research, needed for development.[11]

No one of the "improvements" listed above can be achieved without appropriate and effective political action. Nor will improvement come without political cooperation between donor and recipient countries no matter what is transferred—food, credit, equipment, technological information, research results, or human resources.

A political device put in place by President John F. Kennedy for a variety of reasons, hunger reduction not the chief among them, is the Peace Corps. This is a small-scale, modestly funded, cost-efficient program that has come alive again in the middle 1980s. When the Peace Corps began in 1961, the average age of a volunteer was twenty-four—enthusiastic, idealistic, inexperienced, unskilled. Today the average age is thirty; eleven percent of the volunteers are over fifty. They bring experience and skills (including language skills) to their overseas posts. Half of the Peace Corps volunteers now at work overseas are in sub-Saharan Africa, and most of them are working with the rural poor in food producing activities. The food production will continue after the volunteer departs. Returned volunteers will carry with them a cultural sensitivity and awareness of the hunger problem that will influence their thinking and decision making for the rest of their lives. What they do with their lives is interesting. "Former volunteers now make up 10 percent of every new Foreign Service class. Nearly 90 percent of new recruits for U.S. AID are former Peace Corps volunteers. There are 100 former volunteers working at the World Bank, 200 in staff jobs on Capitol Hill, 14 are vice presidents of Chase Manhattan Bank."[12] Expansion of this U.S. political mechanism would help us to better address the challenge of hunger. Extension of this idea to other "have" nations for the benefit of the "have nots" would be a sure sign of progress in both the promotion and protection of the right to food.

No discussion of political means can ignore the fact that politics is people—at home or overseas. An excellent way to prepare people for international cooperation on the food issue would be to support massive programs of student exchange from the agricultural areas and schools of developing and developed countries. This strategy would focus on the young, before they become policymakers or researchers. They would be exposed to the places where the hunger problem is most acute and where the knowledge leading toward a solution is most advanced. Genuine international cooperation would bring together young people who are only notionally aware that they coexist in a hungry world. Some from the developing nations would not "fit" in the agricultural schools of developed countries, but appropriate "apprenticeship" programs could be designed to expose them to potentially helpful ideas. And while some students from the most advanced schools would find themselves overseas, removed from the best laboratories and libraries for a semester or more, they would gain a new appreciation of the limits to agricultural development in poor lands and the value of experimental stations in areas where the hunger problem is most pressing.

When the political will emerges, political ways will be discovered. The important thing—from the perspective of protecting the right to food—is to never stop trying to cultivate the political will.

The political ways—or means—must be directed, of course, to the appropriate strategic objectives. Under the title of *Feeding the World's Population: Development in the Decade Following the World Food Conference of 1974*, the Congressional Research Service of the Library of Congress produced a 779-page report documenting a decade of progress in world food production (except in Africa) and the presence of "intractable" problems of food distribution worldwide. "Clearly," the report stated, "problems of distribution—of family income, foreign exchange, nutritional knowledge, storage and transportation facilities, and relief programs—are more important to the solution of the world's hunger problem, and also more intractable, than the World Food Conference had foreseen."[13]

The interest of the small and relatively obscure Select Committee on Hunger of the U.S. House of Representatives, created in 1984, is "to determine in what ways the foreign assistance programs of this country can more effectively address the chronic hunger and malnutrition of the people who reside in the nations that are recipients of U.S. foreign assistance." In a 1986 study prepared by the Congressional Research Service for the Select Committee on Hunger,[14] various foreign assistance programs are grouped into major types of activities. These groupings represent the contents of our national political "tool-kit" for the task of hunger education overseas. The four major groupings (hence the four principal tools) are: development assistance, food aid, economic support fund, and military aid. In 1986, development assistance received 27.3 percent of our foreign aid expenditure; food aid, 8.5 percent; the economic support fund, 24.6 percent; and the lion's share—39.6 percent—went to military aid (see Table 1).

The regional allocation of U.S. foreign aid is skewed in directions that reflect our relatively heavy interest in military aid and the economic support fund (a program with foreign policy objectives closely aligned with U.S. military interests) as compared to our investment in regions where hunger is most pressing and our military and security interests are more remote (see Table 2).

The combined development assistance and food aid categories represent just about the sum total of the tools we have to deal with the distribution problems cited above in the status report on the worldwide situation ten years after the 1974 World Food Conference. Relief programs, storage facilities, transportation networks, nutritional education, improvements in family income, and foreign exchange—these goals are far more likely to be met through development and food aid programs (e.g., AID bilateral functional development accounts; AID operating expenses; AID miscellaneous programs like disaster aid; the Peace Corps, Inter-American Foundation, African Development Foundation, Trade and Development; contributions to multilateral development banks and international organizations; and the P.L. 480 program), than through the economic support fund and the four major forms of military aid: the military assistance program (MAP), foreign military sales (FMS) credits, military training (IMET), and peacekeeping operations. The redirection of allocations by region and the shifting of appropriations from one category to another are matters of political decision. In a representative democracy like ours, citizen education related to these complex issues is a prerequisite to citizen action targeted on policy formation and the political decision-making process.

Consideration of political means suitable to the task of eliminating world hunger must transcend the politics of any single nation state and take a global perspective. Proposals, policies, and projects relating to

Table 1.

Composition of U.S. foreign aid appropriations, 1977–86

In millions of current dollars

	1977	1978	1979	1980	1981	1982	1983	1984	1985	1986
Development assistance	2,487	2,781	3,963	3,710	3,559	3,941	4,302	4,233	4,779	4,147
Food aid	1,169	923	806	886	1,229	1,000	1,028	1,227	1,355	1,299
Economic support fund	1,735	2,202	1,922	2,007	2,025	3,065	2,993	3,302	3,902	3,741
Military aid	2,022	2,509	2,981	2,058	3,185	4,104	5,536	6,480	5,910	6,027
Total	7,413	8,415	9,672	8,661	9,998	12,110	13,859	15,241	15,946	15,214

Percent of total appropriation

	1977	1978	1979	1980	1981	1982	1983	1984	1985	1986
Development assistance	33.5	33.0	41.0	42.8	35.6	32.5	31.0	27.8	30.0	27.3
Food aid	15.8	11.0	8.3	10.2	12.3	8.3	7.4	8.1	8.5	8.5
Economic support fund	23.4	26.2	19.9	23.2	20.3	25.3	21.6	21.7	24.5	24.6
Military aid	27.3	29.8	30.8	23.8	31.9	33.9	39.9	42.5	37.1	39.6
Total	100.0	100.0	100.0	100.0	100.0	100.0	100.0	100.0	100.0	100.0

Source: Select Committee on Hunger, U.S. House of Representatives (November, 1986)

Table 2.

Regional allocation of U.S. aid: 1977–86

In millions of constant dollars

	1977	1978	1979	1980	1981	1982	1983	1984	1985	1986
Latin America	689	750	715	711	834	1,149	1,474	1,686	2,300	1,714
Asia	2,209	2,235	1,878	1,656	1,564	1,528	1,864	1,995	2,100	1,840
Middle East	5,488	5,311	8,014	4,394	5,043	5,307	6,029	6,054	5,468	5,475
Europe	1,078	1,472	1,027	1,115	1,131	1,536	1,745	2,106	2,170	1,843
Africa	724	936	849	1,122	1,134	1,236	1,149	1,212	1,236	919
Total assistance	10,187	10,703	12,483	8,998	9,706	10,756	12,261	13,053	13,274	11,791

In percent

	1977	1978	1979	1980	1981	1982	1983	1984	1985	1986
Latin America	6.8	7.0	5.7	7.9	8.6	10.7	12.0	12.9	17.3	14.5
Asia	21.7	20.9	15.0	18.4	16.1	14.2	15.2	15.3	15.8	15.6
Middle East	53.9	49.6	64.2	48.8	52.0	49.3	49.2	46.4	41.2	46.4
Europe	10.6	13.8	8.2	12.4	11.7	14.3	14.2	16.1	16.3	15.6
Africa	7.1	8.7	6.8	12.5	11.7	11.5	9.4	9.3	9.3	7.8
Total assistance	100.0	100.0	100.0	100.0	100.0	100.0	100.0	100.0	100.0	100.0

Source: Select Committee on Hunger, U.S. House of Representatives (November, 1986)

the vast complex of monetary and financial relationships between and among nation states are the raw materials of what some envision as a New International Economic Order (NIEO). The NIEO debate, even in summary form, would carry this essay far beyond its intended limits. Regardless of one's position on the adequacy of present economic arrangements worldwide (recall the principles of human dignity, association, participation, subsidiarity, and preference for the poor), the possibility of a new or renewed international economic order points to the worldwide dimensions of the political arena within which issues relating to the right to food must be resolved. In order to participate in the resolution of these issues, each nation state would do well to attend within its own borders to the cultivation of intercultural sensitivity, linguistic capability, scientific and technical competence, and sufficient political will to end hunger in the world. A universal human right in the economic order, namely, the right to food, requires nothing less.

The right to food establishes a claim to a nutritionally adequate diet for any person anywhere. With that right implied, a May 1987 policy paper drafted to guide the lobbying efforts of Bread for the World stated that the primary objectives of U.S. Agricultural policy should be to:

1. Assume national food security.
2. Help achieve world food security.
3. Help ensure fair returns to farm operators and workers.
4. Ensure conservation and sustainable use of our resource base.

The criteria by which the BFW statement would judge the acceptability of particular policies relative to these objectives are:

1. Assured access to nutritious foods by all persons (the right to food).
2. Prudent use of resources.
3. Fair distribution of economic rewards and power.
4. Economically viable production and distribution systems.
5. Consistency between agricultural and related public policies.

The statement elaborates the right-to-food criterion as follows:

> Assured access to food, on the national level, requires a stable supply of nutritious food at equitable prices, an effective distribution system, maximum opportunity to earn a livelihood, and food subsidies for those unable to purchase food with their own resources. Internationally, it means trade policies which help assure fairness and price stability, as well as food self-reliance, development programs designed to increase food production in food deficit areas, increased family incomes and improved food distribution, and food aid that responds to need efficiently and effectively without inhibiting agricultural and economic development.[15]

That one paragraph contains a complete U.S. political agenda, domestic and foreign, for pursuit, protection, and promotion of the right to food.

It also defines an area of serious ethical responsibility in the face of world hunger.

Notes

1. Washington, D.C.: National Conference of Catholic Bishops, 1986.
2. Pope Pius XI, *Quadragesimo Anno*, (On Reconstructing the Social Order), 15 May 1931, no. 79.
3. House: H. Con. Res. 393; Senate: S. Con. Res. 66. See *Congressional Record*, 25 September 1975.
4. U.S. House of Representatives, Committee on International Relations, *The Right-to-Food Resolution, Hearings before the Subcommittee on International Resources, Food, and Energy*, Ninety-Fourth Congress, Second Session on H. Con. Res. 393 (Washington, D.C.: U.S. Government Printing Office, 1976), p. 257.
5. New York: Paulist Press, 1982.
6. See John Paul II, *Puebla: A Pilgrimage of Faith* (Boston: St. Paul Editions, 1979), p. 148. This is a compilation of speeches taken from the English edition of *L'Osservatore Romano*.
7. For a full text of the Kissinger speech see *War on Hunger*, A Report from the Agency for International Development, vol. 8, no. 12 (December 1974); the portion cited appears on p. 24.
8. Ibid.
9. Baltimore: Johns Hopkins University Press, 1978.
10. Ibid., p. 311.
11. See National Committee for World Food Day, *Food and Poverty: Perspectives, Policies, Prospect*, A Study/Action Packet (Washington, 16 October 1985).
12. *The Washington Post*, 2 August 1987, p. A-20.
13. U.S. House of Representatives, Committee on Foreign Affairs, *Feeding the World's Population* (Washington, D.C.: Government Printing Office, 1984), pp. 1-2.
14. *Trends in Foreign Aid*, 1977-86, (Washington, D.C.: Government Printing Office, November, 1986), 21 pp.
15. "Policy Statement on U.S. Agriculture," *Bread for the World*, 802 Rhode Island Avenue NE, Washington, D.C. 20018 (May 1987), 8 pp.

Summary of Discussions

Session I, Following Paper by Father Byron

DISCUSSANTS

C. Gopalan, Director General, The Nutrition Foundation of India, New Delhi, India

Robert W. Kates, Director, World Hunger Program, Brown University, Providence, Rhode Island

J. E. Dutra DeOliveira, Faculty of Medicine, Sao Paolo, Brazil

Dr. Gopalan, in his opening response, addressed the practical implications of the basic human rights referred to by Byron. The last of these rights, he asserted, is the key. "The right to food, and the right to human dignity are meaningless," Gopalan maintained, "unless we recognize the right of human beings to achieve levels of social and economic development enabling them to command access to the food that they need and to which they are entitled." Our policies, he suggested, should be directed toward removing the hurdles that stand in the way of individuals and communities achieving those levels of development." While we all recognize the problems of malnutrition, he said, it is important not to limit our efforts to the relief of hunger, starvation, and famine, as these are but the superficial symptoms of larger and more basic issues.

The response of the international community and of national governments to these issues, Gopalan insisted, has been less than adequate. Certainly, international organizations have been generous in sharing the surplus food of affluent, developed nations. "But they have been not quite so enthusiastic when it comes to the question of dismantling those unfair institutional and protectionist policies that underly the inequitable social and economic order, and which contribute increasingly to the widening gap between developed and developing countries." Nor are developing countries blameless, he emphasized. Even where "phenomenal successes" have been achieved in increasing food production, malnutrition still exists. Corresponding efforts have not been made to ensure equitable distribution of this food. The "populist solution," observed Gopalan, has been to sweep the problem of poverty under the carpet

William J. Byron, S.J.; Jessica T. Mathews; C. Gopalan; Robert W. Kates; J. E. Dutra DeOliveira

through extensive welfare and feeding programs that serve only to tide individuals over for the time being. These programs don't address the basic problems, which must be solved by improving education, vocational training, and economic opportunity.

Dr. Kates began his opening response with what he termed "reflections on Father Byron's reflections." In connection with Byron's caution against an individualistic approach to human rights, Kates ventured to add a further dimension that he felt characterized liberalism in the Western world: a concern for political and civil rights, and a "failure to address the silent violence of hunger, poverty, and unemployment."

Kates discerned an understandable tendency at such gatherings as this to assume something of a self-congratulatory air. In honoring Dr. Swaminathan, participants might perhaps bask a bit in his reflected light. It was therefore important, Kates felt, to have Byron remind the participants, particularly the Americans, of the hollowness of some of the U.S.'s national rhetoric. Kates found it "deeply troubling," for example, to be reminded of the American government's stated 1974-75 national objectives to make the elimination of hunger a centerpiece of U.S. international assistance efforts. Since that time, he noted, American development and food aid per person (in constant dollar value) has been reduced by about a third.

Kates concluded by noting a "small but potentially powerful initiative in focusing human rights concerns on hunger." Over time, he said, mankind has expanded its notion of whom it defines as human, or as similar to itself. Kates was encouraged to find recent indications that the expansion of this notion is reaching a critical point. "I prefer to think of them as the harbingers of a popular consensus to end the hunger, to see hungry Africans or landless South Asians as an extension of our kind; a consensus and an extension that must underlie all assertions of universal rights." To free the world from hunger, now and in the future, Kates believes that we will need the best of our science, not merely in the obvious applications of technology and productivity but, even more importantly, in the social understanding of how to increase productivity without further burdening the needy. "We also need to know how to sustain agriculture and distribute its products in that crowded, warmer world toward which we move." As a starting point, Kates proposed that "we might begin to speak out when people, particularly civilian populations, are intentionally deprived of food, usually in the midst of conflict, held as hostage to their hunger, pressed for an advantage, punished for their allegiance."

Kates insisted that the end of famine (as opposed to the more tenacious problem of chronic hunger) is well within our grasp. "We now have the productivity to declare that famine is never an act of God, and we now know much about how to make food accessible to those in need." In wartime, however, this knowledge fails us. "To make real the Geneva Convention, to label the use of hunger against civilian populations as the war crime that it is, is a small but important step. In so doing, we can bring together the dual sides of human rights and, with famines demised, create a milestone in the march of our common humanity."

Father Byron responded that, in the last dozen years, he has noticed signs of a development in colleges across the country that may signal a bit of a cultural shift, or "cultural contradiction." The "best and brightest" students of this generation, Byron reported, are experiencing extraordinary anxiety about their abilities to gain meaningful employment. In a world of mammoth need, Byron believes, there is great potential for meaningful work by these young people, and he has seen isolated cases of youth abandoning lives of affluence to seek, in lands of poverty, the kinds of meaningful employment they cannot find at home.

Dr. DeOliveira's attention was drawn most immediately in Father Byron's paper to the reference to protection of the poor. From an ethical point of view, he asked, what should be done when a person in a developing country works hard daily, yet does not earn a sufficient amount of money to buy needed food for himself and his family? What would happen in this situation, DeOliveira wondered, in the United States or in Europe? "I'm not speaking of acute hunger, but of chronic hunger that affects millions of people everyday. How is the principle of human dignity considered by international programs of free food donation and distribution, in evaluating the food needs of poor people who work? Is it ethical and acceptable to solve world hunger through charity? Is it possible that this system might contribute in some way to the perpetuation of world hunger rather than to the solving of it?"

Food and nutrition problems, DeOliveira insisted, cannot be solved in isolation from national and international economic and political conditions. In Brazil, for example, DeOliveira noted that authorities are faced with a choice of producing more cash crops for export, in order to pay off foreign debts, or producing vital foodstuffs that can be sold to a hungry populace only at a discount. "Which is the ethical solution?" he asked. DeOliveira asserted that the world community must remain mindful of the principle of human dignity, for dignity is plainly being denied to individuals and groups who do not have enough food to eat on a daily basis. People who now have access to food, he insisted, "should not forget those who are not eating every day."

Father Byron expressed basic agreement with DeOliveira. The dilemma posed by cash-cropping, or export cropping, versus production for domestic consumption in countries such as Brazil, is clearly an ethical issue, Byron conceded. It is also a problem in which those in the developed countries actively participate. Much of the produce on the tables in Washington, D.C., for example, comes from countries in the South. In developing countries, Byron suggested, the ethical issue would concern who controls the agricultural land and who makes the production decisions. In developed countries, consumers make certain decisions in the supermarket that create the market for the cash-cropping in developing countries. Those countries would surely be better off, declared Byron, if the food they produced for developed nations' stores was devoted instead to their own chronically undernourished populations.

From the floor came an inquiry about why the U.S. has apparently regressed in its commitment to food aid during the past decade, at a time when consciousness of the problem seems to have been raised. The basic problem of hunger, after all, was identified as lack of income as far back as 1850, in an East Africa Company study.

Father Byron cautioned that societal and economic conditions have changed dramatically in the last hundred years, though these changes themselves are illuminating in considering the problem of hunger. In pre-industrial society, he observed, many people had virtually no income, but they were not hungry because they could produce food for themselves. In today's interdependent world economy, money assumes a greater importance for the individual, as it carries with it the ability to buy food that most people cannot produce for themselves.

Byron considered that a strategic mistake was made in the 1950s by planners—even by those within the developing countries—in placing undue emphasis on industrial development while neglecting the importance of agricultural development. Over the past ten years, Byron indicated that he had seen some rectification to that situation, but he regretted that it would take a long time for results to make themselves felt. "We've got to take agricultural development in the LDCs much more seriously," maintained Byron. "We've got to develop a good deal more respect for the small farmer than we've given him in the past."

The real issue, an unidentified speaker from the floor insisted, is not the amount of money the U.S. spends in development and food aid. The focus of international support to developing countries should not be on giving away food, but rather on devising ways to help developing countries produce food. Some equitable economic help might come in the form of higher payments for agricultural or raw products from developing countries. Currently, these goods are sold to industrialized nations at relatively low prices, and manufactured goods are bought back by developing countries at relatively high prices. Ultimately, the speaker insisted, the parts of the world community will have to work together as partners. Developing countries should not simply send people to the U.S. or Europe for advice and assistance. Rather, it is essential that people from developed countries work in less developed countries, becoming familiar with local conditions, and help to increase production there, not only of cash crops, but of basic foodstuffs for local consumption.

Comments from Patricia Kutzner concerned the dynamics of hunger in the U.S. The same kind of dynamics, she believed, accounts for hunger in many other countries. "Our national resources, production, and wealth are trickling very rapidly to the top, rather than the bottom." Echoing Byron's and Gopalan's concerns, Kutzner insisted that the failure to produce more egalitarian income opportunities is at the root of our failure to address the hunger problem. "This whole issue of what are economies geared to do," she claimed, "is the central problem of hunger." Referring to the national debts mentioned by DeOliveira as burdening such countries as Brazil, Kutzner declared that they are "staggering obstacles to efforts to feed people in those countries, compelling them to produce goods for export rather than for their own people's needs." Developed nations, she pointed out, share responsibility for the economic planning of bygone years that produced these debts.

L. K. Bharadwaj expressed strong support for Father Byron's proposal for a massive student exchange program between developed and developing Third World countries. Bharadwaj pointed out that in 1976, M. S. Swaminathan conceived a unique proposal for a rural reserve corps of exchange professionals. The scheme called for a cluster of experts backed by a consortium of scientific institutions. It is high time, suggested Bharadwaj, that this idea be implemented on a large scale.

Jim Phippard pursued the theme of "political will," as raised in

Father Byron's paper. Phippard cited a striking illustration of the differences of political will in the differing reactions elicited by famine situations (as opposed to hunger situations that do not approach the level of famine). During the Ethiopian famine of 1984-85, noted Phippard, there was a tremendous outpouring of feeling and money from the U.S. and other countries, largely in response to the media's graphic illustrations of the problem. In contrast, today we are facing drought in Ethiopia, and the strategy for dealing with it is more realistic. An attempt is being made to get people to stay in their own villages and continue living in their communities, rather than going to huge refugee centers, where there would inevitably be tremendous health problems. The resulting absence of any graphic images of famine crisis in the media has, however, presented difficulties in stirring public support for such organizations as the Catholic Relief Service, which recently mounted a media campaign focusing not on people dying, but on the urgent efforts to increase agricultural production. The CRS campaign was not a success. For similar reasons, there has been a problem marshaling political support for essential foreign aid programs on Capitol Hill. Politicians' constituents will react to famine, but they tend to think of foreign aid as a "giveaway" program. Phippard believes the question, therefore, is how to get the grass roots understanding of, and political support for, the requirements of international development beyond the essentials of famine relief.

Father Byron reflected that there probably was no simple answer to the question. Politics ultimately consists of nothing but people, however, and the closer one can focus decision makers' perceptions on areas of actual crisis or chronic malnutrition, the greater will be the possibility of opening up human emotions that could translate into political will.

SESSION II

The Innovative Environment for Increased Food Production in Africa

The dramatic events of the 1983-86 drought and hunger in Africa demonstrated more clearly than ever the contrasting successes experienced by Mexico, India, Pakistan, and several other countries in Asia and Latin America in transforming their agricultural production to food self-sufficiency in the last twenty years through the development and adoption of Green Revolution technologies. This particular technological approach has not been inherited by Africa in any substantial way. The Green Revolution ushered in a new agricultural revolution in the tropics and subtropics and, within a matter of years, has transformed the wretched basket case that India was before 1966 into a grain-surplus country. In wheat, rice, and maize, we see many tropical developing countries joining the great farming economies in North America, Europe, and Australasia as green-revolutionized, food-rich countries, although not necessarily well fed at all strata of the society. But Africa lags behind. Over the same time period, 1966 to 1986, it has shown a decided trend toward domestic food shortages, sometimes breaking into famine in drought or flood years. For instance, Nigeria's food exports accounted for more than 70 percent of its GNP at independence. A quarter of a century later, food items accounted for more than 50 percent of its annual imports (Obasanjo, *Issues in Agricultural Development in Africa*, 1987).

What has made Africa such an exception? What can we do to spur on a different type of agricultural revolution to fit the specific circumstances and the special environment of Africa?

Thomas R. Odhiambo
Director
THE INTERNATIONAL
CENTRE OF INSECT
PHYSIOLOGY AND
ECOLOGY

The African Challenge to Science

There is no doubt that a "green advance" seems a realistic hope in Africa, even though this advance may not necessarily take the form the Green Revolution of Asia and Latin America has taken. The agricultural challenge in Africa is a profoundly difficult one, as it should focus on the resource-poor farming households, who produce 70-85 percent of the food in this region. Agricultural scientists must therefore wear a different set of intellectual spectacles in order to appreciate the special demands of the technological terrain—one in which they should be operating with the active participation of this client rural community.

The fact that the preponderant farming community in Africa is the resource-poor, small-scale, usually subsistence farmer will fully stretch our intellectual and scientific capabilities if we are to reverse the present trends in food production and start on a course of sustainable food self-sufficiency.

Firstly, almost every problem in agricultural production is more acute in Africa than elsewhere. The African tropical soils are more fragile, and they are anything but the fertile, flat bottomlands that made the Punjab the crucible for the start of the Green Revolution in Asia. The shortage of trained agricultural scientists is acute, the performance of extension officers is often indifferent, and the institutional framework for scientific research, technological development, and farmer relations is weaker and generally unresponsive to the operational requirements. Except in the Sudan and Egypt, there is little technological culture in irrigation, and most agriculture is therefore wholly dependent on the vagaries and uncertainties of rainfall. Outbreaks of migrant pests—locusts, armyworms, quelea birds, and the like—have become a recurrent theme of the woes of the continent. The safety security net that quarantine should provide against exotic diseases and pests of the major crops and livestock is largely ineffective. The agricultural market, which forms 30–58 percent of the GNP of African countries, has been falling for some time.

Secondly, it is not possible to conceive a long-term, sustainable agricultural production system for the African resource-poor farming households without effecting a rural health revolution. Malaria, bilharzia, river blindness, filariasis, and other vector-borne diseases have kept African wetlands from being intensively developed. Moreover, leishmaniasis, sleeping sickness, and diarrheal diseases of the young keep the vast semiarid areas of Africa sparsely populated and development poor. Similarly, the diseases of African livestock are equally horrendous (see Table 1). These keep vast areas of Africa inhabited only by small, tough animals that are genetically able to withstand the ravages of diseases and pests to some considerable extent, but not enough to make a reasonable yield from them. Policymakers and planners in Africa have hardly recognized the fact that, in programming an agricultural revolution in most of Africa, they must simultaneously assure an integrated rural health program in order to create the necessary incentive to grow out of mere rural survival and subsistence.

Thirdly, development—particularly agricultural development, which is so circumscribed by the cultural, socioeconomic, and biological universes of the particular people and region we are targeting—cannot be imported, because the technology on which it is based cannot effectively be imported. The relevant technology for African tropical agriculture can only become sustainable if it is anchored in the cultural-societal milieu of the region, and within the scientific framework resulting when relevant questions are asked by the farming community. As former Head of State of Nigeria-turned-farmer, General Olusegun Obasanjo, states:

> Technology, which apparently drives the world of the late twentieth century, is the produce of science. It is, in fact, applied science, and cannot be developed without a science base. The transfer of technology is an illusion, a catchy phrase conjuring up images of high-level scientific and technical expertise willingly and altruistically handed over in gift-wrapping from the owner in the developed world to us in Africa and the rest of the Third World. (Obasanjo, *Africa in Perspective*, 1987)

The poignant challenge for Africa, then, is to develop its human capital to a critical mass. It must be motivated to find and implement long-term

Table 1.

The economic and social impact of animal diseases in Africa

Type of Impact	Principal Diseases
Mortality	Rinderpest, tick-borne diseases, trypanosomiasis
Lowered Productivity	
(a) Infertility	Malnutrition, brucillosis
(b) Loss of weight	Malnutrition, distomatosis, gastrointestinal parasites, ticks
(c) Loss of draft power	Malnutrition, distomatosis, gastrointestinal parasites, ticks
Denial of Land Use	Trypanosomiasis
Denial of Export Markets	Rinderpest, foot-and-mouth, contagious bovine pleuropneumonia
Impairment of Genetic Improvement Programs	Tick-borne diseases, streptothricosis, foot-and-mouth

Source: Tropag Consultants Ltd. (1986), modified

solutions to its food production, marketing, and security problems in complete concert with its farming community.

Fourthly, and finally, the transition in Africa from today's traditional agricultural practices (usually implying subsistence), to modern, science-driven agriculture (associated with employment, as well as with economic and food security for resource-poor households) must start from an environment of innovation. This can be created and nurtured only by Africa's geopolitical leadership, which can provide the necessary social space for an agricultural revolution to be initiated and maintained in the continent. The transition will require the rationalization of the traditional knowledge base, upgrading it to modern circumstances, and fusing it with novel ideas and practices.

The one major gain produced by the Green Revolution in India is that it instilled a sense of self-confidence in many developing regions that they too have the capability to achieve food self-sufficiency. As M. S. Swaminathan, a pioneer player in the Green Revolution technology, so succinctly states it:

The Green Revolution in India destroyed the myth that agricultural evolution in developing countries has to be a very slow process and that revolutions in crop production are not easy because of the very large number of small-scale farmers whose active involvement is essential to make such revolutions possible (Swaminathan 1987).

The question posed by resource-poor farming households in tropical Africa is how to provide low-input, low-cost technology that is appropriate to their conditions of resource poverty. In dealing with this question, fundamental research and technological development efforts must also consider the special agro-ecological conditions, both biological and socioeconomic, imposed by the African environment.

Technology For African Agricultural Transformation

Tropical Africa traditionally practices a highly complex cropping system. Several crop species, quite often including perennial crops, are planted in the same time and space. This system provides food, condiments, and domestic fuel, and gives some insurance against crop failure due to the vagaries of weather, pest and disease outbreaks, and occasional mistiming in planting. Such a methodology involving intercropping and agroforestry has only recently been recognized by agricultural scientists (both in temperate regions and in the tropics) as an ecologically effective means of managing sustainable crop production in areas with a great biological diversity, rain-fed agriculture, and high pressure from pests and diseases. This methodology gives a greater total yield per unit area than if the same crop components were grown separately as sole crops, even when these are cultivated with methods of modern technology (Okigbo 1978). The factors responsible for this higher yield are complex, but they include the more efficient utilization of biological resources by the differing rooting systems and shoot architecture; the dramatic drop in the incidence of plant diseases (Harwood 1976); and the suppression of insect pest populations under certain crop mixtures (Odhiambo 1984). The success of this methodology is demonstrated in a comparative study conducted in western Kenya of the productivity of cropping patterns in terms of gross returns per hectare expressed as land equivalent ratios (LER). Amoako-Atta and Omolo found that a sorghum-cowpea-maize tricrop commands a 1.65 LER, and that a sorghum-cowpea dicrop results in a 1.30 LER, in comparison to monocrop ratios of only 1.0 (Amoako-Atta and Omolo 1983). Similarly, alley cropping (an upgraded practice arising from traditional fallow and agroforestry) can effectively prevent run-off and soil erosion from the fragile soils of lowland humid tropics, while still maintaining the same level of crop production (see Table 2) (Lal 1987).

Table 2.

Effects of alley cropping on run-off and soil erosion under maize-cowpea rotation, at Ibadan (Nigeria), 1984

Treatment	Run-off	Soil Erosion	Crop Production	
			(ton/ha)	
	(mm)	(ton/ha/yr)	Maize	Cowpea
Ploughed	232	14.9	4.2	0.5
Zero Cultivation	6	0.03	4.3	1.1
Leucaena	10	0.2	3.9	0.6
Gliricidia	20	1.7	4.0	0.7

Source: R. Lal (1967), modified

Initial steps have been taken in the systematic investigation of the ecological basis for the success of intercropping and agroforestry under certain farming conditions. In the future, we should expect to go further than simply understanding the system in scientific terms.

We should begin to develop farming tools, for upgraded intercropping and agroforestry systems, that minimize backbreaking labor. This is particularly important in farming communities that are now beginning to feel a stressful shortage of labor at critical times in the farming cycle—during weeding, pruning, and harvesting. The African hoe, for example, is still more or less what it was some two thousand years ago. The modern high-energy tractor, on the other hand, was designed for large-scale farming routines, and is irrelevant to the needs of the small-scale farmer, quite aside from the fact that its acquisition is far beyond the reach of the resource-poor farmer.

Research is the art of asking the critical question and developing the appropriate methodology for answering it. We should now set ourselves on the right path of African agricultural research, by bringing in the resource-poor farming household to jointly establish the research and development agenda.

Part of that research and development agenda must comprise the major staples of tropical Africa, which have received such scant scientific attention, yet provide a very large component of the food resources of the continent. Of the ten major staples of East and Southern Africa, for instance, two (the sweet potato and banana) are barely known in agronomic and scientific terms, and three others (cassava, sorghum, and millet) have only recently acquired a legitimacy for scientific inquiry—even though their contribution to food self-sufficiency at the rural household level is crucial and preponderant (see Table 3). For instance, the banana is a major staple in Burundi and Rwanda, occupying some 24 percent and 42 percent of the cropland, respectively. In Uganda it is credited with having saved the country from stack famine in the ongoing fifteen-year civil unrest. Yet, in a banana-dependent population of some nine million, only five research scientists in the whole Central African region are thought to be involved in any significant manner on banana research. We know almost nothing about sweet potatoes—not even the level of their consumption in Africa. Nor do we possess any significant scientific details about the germplasm we possess. We do know that zero grazing of goats in Uganda (in the Ukerewe Islands) and in Tanzania (in the Kilimanjaro area) is dependent on the sweet potato. Still, we have yet to examine this traditional practice in modern scientific terms and to investigate the special agronomic requirements that make it productive. Even in the case of sorghum and millet, which have received some scientific attention in recent years, there are no on-shelf technologies that would meet the needs of the rural households for both grain and fodder (Kyomo 1986).

Equally dismaying is the matter of livestock production, since most of the livestock keepers in Africa turn out to be, in reality, subsistence livestock owners (Tropag Consultants 1986). Eastern and Southern Africa contain approximately 108 million cattle, or about 8.5 percent of the world's cattle population. The bulk of the research and development activity in this region is targeted to cattle. On the other hand, this region also holds: 9.9 million camels (or approximately 57.6 percent of the world's camel population); 75 million goats (or about 16.3 percent of the world's goat population); and 5.2 million donkeys (or just over 13.1 percent of the world's donkey population). Yet, the camel and the donkey are almost entirely ignored by efforts to upgrade technology for improved livestock productivity. The African rural and pastoral households who keep them rely entirely on their traditional knowledge base for their breeding, husbandry, utilization, and marketing. Though cherished by the mixed farmer and pastoralist alike, the goat has, in the meantime, acquired an undeserved notoriety as a major cause of desertification in many parts of Africa. The controversy over this matter has obscured the great potential of the goat for the resource-poor farmer and animal husbandman. As French so enthusiastically puts the case:

> Because of their small size and clean habits, goats can be kept in areas and in quarters which would be unacceptable for cows. They can profitably utilize underdeveloped or degenerated land on which cows would starve. They are patient, attractive and affectionate animals which are easy to

Table 3.

The ten major staples of East and Southern Africa

Staples	Countries Where This Staple Occupies More Than 20% of the Total Staple Crop Area
Potato	Mauritius
Sweet Potato	None (Burundi the highest, with 13%)
Banana	Uganda, Burundi, Mauritius, Rwanda
Cassava	Tanzania, Mozambique, Zaire
Sorghum	Burundi, Rwanda, Lesotho, Botswana, Sudan, Somalia
Millet	Zimbabwe, Uganda, Sudan
Rice	Mauritius, Madagascar
Wheat	Lesotho
Barley	None (Ethiopia the highest, with 17%)
Maize	Mauritius, Somalia, Uganda, Burundi, Zaire, Tanzania, Mozambique, Lesotho, Zimbabwe, Angola, Kenya, Zambia, Malawi, Swaziland

> feed, manage and maintain. Consequently, they constitute valuable domestic animals and profitable commercial milk producers, and in both cases, pay for themselves as well as for the feed supplies and the labour involved in collecting and handling their milk. . . . This indifference to their attributes and commercial possibilities is rather surprising (French 1970).

Goats were domesticated at least seven thousand to eight thousand years ago—long before the domestication of sheep, cattle, and pigs. From pictorial records, we know that Africa was one of the regions where the goat first appeared as a domesticated animal in prehistoric settlements. Africa should surely be a leader in research and development in goat production, as it should be in camel and donkey improvement.

Appreciating what we know about the agro-ecological and socio-economic environment of tropical Africa, the challenge for the scientific community now is how to expand the individual farmer's opportunities in fundamental ways. What feasible technologies can experimental science develop to respond to the real needs of the resource-poor farming households? How can we create the potential to free them from dire deprivation, so as to release their entrepreneurial resources and human capital, which are now tied up in the consuming search for mere survival?

It is now generally realized that a major cause of famine and malnutrition is poverty. Consequently, if the agricultural research systems, both national and international, have tended to concentrate on the agricultural production aspects in considering this issue, it is a reflection of their disciplinary bias rather than a reflection of the practical situation. It is therefore refreshing to find a recent international conference in Rome organized by the Accademia Nazionale delle Scienze detta dei XL conclude that "in the next phase of the Green Revolution, as much attention will have to be given to fighting the famine of jobs as to working for adequate food supplies, nutritional level and protection from disease" (Accademia 1986). Thus, there must be as much concern devoted to the economic sustainability of the farming households of tropical Africa as there is for the ecological sustainability of the new agricultural technologies.

Often, one is confronted with disturbing statistics regarding African population growth in relation to development strategies, including the question of long-term food security. In examining these, one needs to remember that health, fertility, and population growth are all intimately related.

It has been demonstrated that improved health is a vital factor in the transition from high child mortality and reproductive fertility rates to lower levels, eventually leading to declining rates of population growth (Measham 1986). With the multitude of health burdens faced by Africa's children from their birth—malaria, respiratory complications, diarrheal diseases, and malnutrition, among other afflictions—child mortality is high, and life expectancy the lowest in the world. We are unlikely to make a lasting dent in the task of making the resource-poor farming household meet its food requirements as well as its basic economic needs unless we accomplish two immediate goals. First, we must develop low-cost, effective technologies, in order to give improved returns on family labor, which is usually the principal input of such farming households (The World Bank, "Strengthening Agricultural Research in Sub-Saharan Africa," 1987). Secondly, a certain level of rural health must be attained, through a mixture of preventive health measures and curative medicine, in order to provide a modicum of relief from the wretchedness of tropical rural life, and thereby provide a climate for considering a more hopeful future—an incentive to do more than just survive.

The motivational level of the research and development community—those directly responsible for implementing the research agenda jointly devised with the farming community—is small (see Table 4). It is, in fact, subcritical in size in the majority of African countries and is ill rewarded and painfully unheralded.

Table 4.

Indigenous research scientists in relation to agricultural population in selected African countries, 1984

Region/Country	Agricultural Population (millions)	No. of Research Scientists Per Million Agricultural Population
Botswana	0.6	28
Lesotho	1.2	5
Zimbabwe	4.7	42
Malawi	5.6	21
Mozambique	7.0	2
Tanzania	15.6	18
East Africa	—	9
West Africa	—	43
Latin America	—	60
Asia	—	16

Source: M. L. Kyomo (1986), modified

The Condition of the Agricultural Research Community in Africa

A great deal has been written about the weakness of the African national agricultural research system: the training and further training of its research leadership, the intellectual environment in which it operates, its budgetary and physical support, the manner in which it establishes its research and development priorities, and its relationship to its clientele—the extension service, the agribusiness community, and eventually the farming households themselves. The World Bank's recently released (1987) report, *Eastern and Southern Africa Agricultural Research Review*, summarizes quite well the status of the agricultural research system in this region. Several aspects are particularly serious:

1. Substantial postgraduate training programs in the agricultural sciences and related studies exist only in a few countries.
2. There is an acute lack of knowledge services or disjointed operation of existing ones: the libraries, journals, newsletters, and external contacts,

which together inform the research and development community, as well as university students, of advances in the agricultural and associated sciences, in their countries as well as abroad.
3. There is a pervasive insufficiency of training at the advanced degree level, and therefore little training in research and analytical methods. Indeed, the research staff is comprised largely of first-degree holders (Kyomo 1986).
4. There is a high attrition rate—something on the order of 7 percent per year, or more. Brain drain has become a worrisome problem. As Kyomo has so dramatically stated: "Although policy makers have supported the notion that scientists should be paid better salaries than those purely in administrative jobs, there has been no implementation of such a policy."
5. There is an evident lack of entrepreneurship in the science enterprise, because of increased bureaucratization of the research service.

The cumulative consequence of these factors is that the national agricultural research system in Africa is the weakest link in the chain—from an agreed research and development agenda, to innovation, and ultimately to the implementation of feasible production and marketing strategies. The situation is made more desperate by the fact that the U.S. spends only $300 million annually on agricultural research in sub-Saharan Africa (Measham 1986), falling short of the 1-2 percent of agricultural GDP that is the minimum essential level, in the estimation of the July 1985 summit meeting of the Heads of State and Government of the Organization of African Unity (OAU), which met to consider a strategic approach to Africa's ongoing economic and food crisis (OAU 1985). Furthermore, the funding for research and development is not assured on a medium- to long-term basis, and is largely geared toward meeting emoluments, thus effectively paralyzing scientific activities themselves.

The agricultural research leadership, in concert with the geopolitical leadership, must create and nurture an environment that encourages research entrepreneurship and invigorates technological innovation. The scientific community needs to be assured of an atmosphere of, and a facility for, easy exchange of scientific information and experience, and an environment of robust peer competitiveness. Scientists feed on recognition for excellent work, although—unlike many other professionals—they must often barely subsist on livings that do not adequately recognize their worth and the risks they take in building up know-how and know-why. In attempting to establish an indigenous critical mass of productive agricultural scientists in Africa, it thus becomes quite clear how vital it is to design and implement an institutional mechanism that will identify and reward outstanding research entrepreneurs:

> The outcome of research is subject to risk and uncertainty, and research scientists, in addition to professional competence, organizational skills, and access to funds, need a special human talent: research entrepreneurship. The ability to articulate research hypotheses and to choose research priorities from among a multitude of possibilities, and then to put up with risk and uncertainty inherent in all research, is an innate and hard-to-define talent. . . . Research leaders in Africa must devise institutional structures and mechanisms to select, encourage, and reward such research entrepreneurs (Special Programme for Agricultural Research 1987).

As no less a personality than ex-President Obasanjo has recently suggested, Africa's geopolitical leadership must in this way organize and harness talent from the national business and intellectual communities, so as to design, through the resourcefulness of these research and development entrepreneurs, agricultural technologies relevant to the farming community (Obasanjo, *Africa in Perspective*, 1987).

We must be clear that what we are seeking to recognize and reward in a special way, in order to create a larger indigenous pool of these talented people, are the innovators. In their celebrated book, *In Search of Excellence*, Peters and Waterman made a distinction between "creativity" (thinking things up) and "innovation" (doing new things). They concluded:

> There is no shortage of creativity or creative people. . . . The shortage is of innovators. All too often, people believe that creativity automatically leads to innovation. It doesn't. . . . The fact that you can put a dozen inexperienced people in a room and conduct a brainstorming session that produces exciting new ideas shows how little relative importance ideas themselves have. . . . The scarce people are the ones who have the know-how, energy, daring, and staying power to implement ideas (Peters and Waterman 1982).

These are the talented people that Africa needs among its research scientists, extension officers, and marketing community.

Prospects

As we survey the agricultural prospects of the African resource-poor households for the next fifteen to twenty years, there is no doubt that new production technologies, probably created by research entrepreneurs steeped in tropical agro-ecology and taking advantage of modern biotechnology, will provide the motive power for the sustainable growth of agriculture throughout the region. However, the successful adoption of these low-cost, low-input technologies will depend on a number of conditions being met. Firstly, that the geopolitical leadership creates an enabling environment for entrepreneurial innovation in research and development, in marketing, and in family employment (including that in nonfarming activities, such as agro-industry). Secondly, that the research and development community takes the farming community into effective partnership in developing the research and development agenda. And thirdly, that the agenda addresses its long-term goals through an initial understanding and rationalization of the traditional knowledge base of Africa's tropical agriculture.

In a gathering of PEN (Poets, Playwrights, Essayists, and Novelists) in New York in 1986, U.S. Secretary of State George P. Schultz is credited with giving this allegorial story:

> Traditional Chinese tales tell of the village papercatcher, the one designated to go about gathering up any stray scrap of paper with writing on it that might be blown along the alleys by the wind—not for the purpose of cleaning up litter, but because writing was precious, and one single piece of paper might be found to contain an enlightening thought which, however fleeting, must not be lost.

One has a feeling that quite a few of the African traditional practices in tree, crop, and animal husbandry may well take the form of the precious Chinese scrap of paper.

References Cited

Accademia Nazionale delle Scienze detta dei XL. 1986. "Towards a Second Green Revolution: From Chemical to New Biological Technologies in Agriculture in the Tropics." Unpublished recommendations of a conference held in Rome.

Amoako-Atta, B., and E. O. Omolo. 1983. "Yield Losses Caused by the Stem/Pod Borer Complex Within Maize-Cowpea-Sorghum Intercropping Systems in Kenya." *Insect Sci. Applic.* 4, no. 1:39-46.

French, M. H. 1970. *Observations on the Goat.* 1970. Rome: Food and Agriculture Organization of the United Nations.

Harwood, R. R. 1976. "The Application of Science and Technology in Long-Range Solutions: Multiple Cropping Potentials." In *Nutrition and Agricultural Development*, edited by N. S. Scrimshaw and M. Behar. New York: Plenum Press.

Kyomo, M. L. 1986. "Agricultural Research in Eastern and Southern Africa: Issues and Priorities." Unpublished report prepared for the Eastern and Southern Africa Agricultural Department. Washington, D.C.: The World Bank.

Lal, R. 1987. "Managing the Soils of Sub-Saharan Africa." *Science* 236:1069-76.

Measham, A. R. 1986. "Health and Development." *Finance and Development* 23, no. 4:26-29.

Obasanjo, O. 1987. *Africa in Perspective: Myths and Realities.* New York: Council on Foreign Relations.

———. 1987. *Issues in Agricultural Development in Africa.* Arturo Tanco Memorial Lecture, April 1987. New York: The Hunger Project.

Odhiambo, T. R. 1984. "International Aspects of Crop Protection: The Needs of Tropical Developing Countries." *Insect Sci. Applic.* 5, no. 2:59-67.

Okigbo, B. N. 1978. *Cropping Systems and Related Research in Africa.* Addis Ababa: Association of Agricultural Sciences in Africa, Occasional Publications Series OT-1.

Organization of African Unity. 1985. *Africa's Priority Programme for Economic Recovery, 1986-1990.* Addis Ababa: Organization of African Unity.

Peters, T. J., and R. H. Waterman, Jr. 1982. *In Search of Excellence: Lessons from America's Best-Run Companies.* New York: Harper and Row.

Special Programme for African Agricultural Research. 1987. *Guidelines for Strengthening National Agricultural Research Systems in Sub-Saharan Africa*, 1-28. Washington, D.C.: The World Bank.

Swaminathan, M. S. 1987. "The Green Revolution." In *The Development of Maize and Wheat in the Third World*, 27-47. Mexico: D.F.: CIMMYT.

Tropag Consultants Ltd. "Livestock Research." 1986. Unpublished report prepared for the Eastern and Southern Africa Agricultural Research Department. Washington, D.C.: The World Bank.

The World Bank. 1987. *Eastern and Southern Africa Agricultural Research Review.* Washington, D.C.: The World Bank.

———. 1987. "Strengthening Agricultural Research in Sub-Saharan Africa: A Proposed Strategy." Unpublished mimeographed paper presented at a high-level seminar on African agricultural research, Feldafing, West Germany, 23-28 September.

Summary of Discussion

Session II, Following Paper By Dr. Odhiambo

DISCUSSANTS

Nyle C. Brady, *Senior Assistant Administrator, Bureau for Science and Technology, U.S. Agency for International Development, Department of State, Washington, D.C.*

Ruben L. Villareal, *Dean, College of Agriculture, University of Philippines at Los Baños*

Mamdouth Gabr, *President, International Union of Nutrition Sciences, Cairo, Egypt*

Dr. Brady opened discussions by underscoring the important interrelationship noted by Odhiambo between agriculture, health (animal and human), population, and hunger. "One of the reasons Africa has had trouble feeding itself," observed Brady, "is that its ability to produce humans has exceeded its ability to produce food."

According to Brady, conditions in Africa are clearly different than elsewhere. Brady concurred with Odhiambo's criticism of the notion that technology can simply be imported to Africa from the U.S., or even from Asia. Methods must be developed in Africa that are appropriate to indigenous conditions. Nevertheless, some basic principles demonstrated in Asia also hold true in Africa. First, insisted Brady, agriculture has to be given higher priority than it has hitherto received from current leaders. In some cases, Brady complained, government actions, such as taxing of exports, have actually tended to discourage agriculture. A second lesson to be learned from Asia is that human resource development must be accelerated at all stages: research, teaching, extension, and even agriculture itself. Third, appropriate technology and technological systems must be developed. The success of hybrid maize in Africa, for example, proved something of a surprise in Africa and elsewhere, because the degree to which so-called native technologies could accommodate a high producing strain was not anticipated. Fourth, changes must be made in public policy that will make it possible for the producer, and those working with him in rural areas, to profit from the new technologies.

The availability of resources such as water and fertilizers, lamented Brady, will continue to pose a problem in Africa's efforts to bring about an

Thomas R. Odhiambo; Jessica T. Mathews; Nyle C. Brady; Ruben L. Villareal; Mamdouth Gabr

agricultural revolution. Hopeful signs have appeared in parts of Africa, where efforts are being made to develop pest-resistant strains of plants that might be used in place of pesticides. Other crops, such as woody species legumes, have been used similarly in place of fertilizers to provide nitrogen for the system. This is all to the good, Brady maintained, because chemicals are frequently more expensive and ecologically deleterious in the long run.

Brady saw the equitable distribution of food as a major problem facing Africa, and one that has not been adequately addressed in Asia. Asian countries, he noted, have done remarkable things with food production, succeeding in some cases in producing surpluses and comfortable reserves for times of drought. But while these enormous food stocks sit in storage, hundreds of millions of people often go hungry.

Brady professed mixed feelings about biotechnology. While it offers tools of great promise, particularly in drought tolerance and pest resistance, he expressed concern that it could produce an even greater gap between the developing and the developed countries. Unless laboratories in First World and Third World countries work in concert on matters of concern to both, developing countries will be placed in an even more dependent position by this new science.

Finally, Brady registered his concern over the dangers posed to natural resources worldwide. It is a challenge to science, said Brady, to develop technological improvements that can, for example, demonstrate to farmers tangible advantages over traditional slash-and-burn techniques.

Dr. Villareal agreed that the story is indeed different in tropical Africa. "Though several countries of Asia and Latin America have dramatically demonstrated the potential for utilization of Green Revolution technologies for raising rapidly high fertilizer and pesticide applications and use of high yield seed in greatly increasing the production of food staples, . . . similar breakthroughs . . . will take longer than in Africa."

Dr. Odhiambo later expressed agreement with Villareal and Brady in this matter. "I think that what this is leading to is a commitment by the scientific community, as well as by the political leadership for a long haul, probably not less than ten or fifteen years, to see these problems through." Odhiambo, in fact, feared that "the stability of the African continent depends on solving this problem, which is not only one of survival, but one of giving the people of that continent, particularly resource-poor farmers, a chance in life, a chance to utilize their own talents, a chance to make a livelihood, and maybe even a chance to make the rural environment more attractive for living than it has been in the past." In Odhiambo's view, therefore, discussions of food also directly concern questions of gainful employment and the development of human capital, particularly as they relate to women.

Considering agriculture in the world economy, Villareal declared that too little attention has been paid to the crucial factor of marketing in effecting agricultural transformation and modernization. He pointed to the successes of agriculture in Taiwan, where marketing experts had thoroughly analyzed the tastes of their export markets, determining such things as what color labels to use for the European Common Market, and what size of cans sell best in Canada, the U.S., or France. "I think this kind of marketing technology and information is somewhat lacking in many of the developing countries."

With respect to Western science and technologies applications, Villareal cautioned against potentially negative long-term consequences. A continuous reliance on Western science and technological inputs, he warned, tends to retard the growth of domestic scientific and technological capabilities. Moreover, Western grants or loans to developing countries often require the purchase of expensive farm machineries and equipment, raising problems of maintenance and efficient utilization.

Villareal also pointed to the hazards of uncontrolled, regular importation of breeding stocks, which tends to exacerbate imbalances of trade exchange between developing and developed countries. It also tends to undermine developing countries' self-reliance, he suggested. In addition to the serious risk of introducing new livestock diseases, such indiscriminate animal stock importation inevitably has the effect of discouraging local breeding farms from developing breeds better adapted to local conditions. "A strong policy to develop local breeding must be adopted if the country hopes to be truly self-sufficient and self-reliant," Villareal insisted.

Villareal described a number of other equally serious problems that he had noticed accompanying the gradual, growing dependence of Third World countries on the import of agricultural technologies and stocks from developed nations. Unless industrialized nations assume responsibilty for assuring the safety of their export products as scrupulously as they monitor their profits, LDCs will face the danger of becoming dumping grounds for harmful materials, surplus goods, and, more recently, genetically engineered vaccines, drugs, and pesticides. In order to regulate the import and domestic sale of these products, governments of importing countries must be given details on the nature of the materials and kept informed about the restrictions to which they are subject in the exporting countries. Industry organizations can also excercise a beneficial self-regulating force on the development and use of pesticides and other products, noted Villareal, citing as an example the Agricultural Pesticide Institute of the Philippines.

Dr. Gabr prefaced his remarks by drawing a distinction between food shortages and undernutrition. Though, as Dr. Odhiambo mentioned, increasing food production is the most important means of combating undernutrition, it is not the only means. Gabr noted that a number of other approaches had been mentioned by Brady and Odhiambo, including the avoidance of food waste, the correcting of maldistribution, the curbing of urbanization and ignorance, the improving of food assimilation and availability, and the controlling of infectious diseases and overpopulation. It might be worthwhile, suggested Gabr, to discuss, from a cost-benefit perspective, the feasibility of overcoming undernutrition through these approaches, rather than relying exclusively on increasing food production. In order to invest in the approach appropriate to the situation, a careful analysis of opportunities for these approaches should be made by each country and, if possible, by each region. For example, said Gabr, Egypt's efforts to combat certain diseases through relatively simple measures such as oral hydration had realized great success in alleviating undernutrition. From a practical standpoint, they may well have been more cost-effective than efforts to increase food production.

Given the focus of the current gathering, Gabr felt it should be kept in mind that the goal in efforts to combat malnutrition is to improve the quality of life, rather than to deal with the larger issue of the right to survival as such. Improving the quality of life through better nutrition cannot be achieved solely by increasing food production, he noted. To achieve this, it will be necessary to correct conditions in several other areas. Among these is the matter of political will, mentioned earlier by Father Byron. Gabr expressed a strong feeling that it is the responsibility of our scientific organizations and unions to bring the issues raised by Odhiambo to the attention of decision makers and the public, so that proper actions can be promoted and implemented. In Gabr's opinion, we are still lamentably short of that goal. Communication strategies, he proposed, should be reassessed and strengthened. Food shortages, with consequent undernutrition, Gabr pointed out, are mainly a problem of developing countries. Unfortunately, he said, many scientific bodies are so deeply involved in their own domestic food and nutrition problems that they fail to recognize this fact. Admittedly, in some instances, politicians in the Third World have exacerbated the problem, actually increasing malnutrition or delaying efforts to combat it. It is therefore the responsibility of scientists "to put them back on the right track."

SESSION III

Food Entitlements and Economic Chains

Hunger, Entitlement, and Linkages

It is not a new question. "Why hunger?" remains as relevant a query today as it has been for thousands of years. But the answer cannot be quite the same as in the past. The enormous expansion of productive power, especially in agriculture, that has taken place over the last few centuries has certainly made it possible to guarantee adequate food for all. The persistence of chronic hunger and severe famine despite more than adequate productive opportunities poses a range of questions that would not have been immediately relevant in the past, when production possibilities were much more limited. This is not to say that issues of technological transformation and expansion of production have ceased to be important. Much more can certainly be achieved, and the rewards from productive expansion can indeed be very substantial. But even with existing technologies and the current state of knowledge, a better marshaling of our resources and a better system of distribution can eliminate much of the nutritional shortfall that is observed across the world today. Hunger and famine have to be seen as economic phenomena in the broadest sense—including production, distribution, and utilization of food—and not just as reflections of problems of food production as such.

Production, distribution, and consumption of food are linked together in the form of chain relationships—connecting producers to markets, markets to purchasers, and purchasers to consumers. Food deprivation, hunger, and starvation can result from dysfunctioning of any of these links. A person suffers from food deprivation and consequently

Amartya K. Sen
*Lamont University
Professor of Economics
and Philosophy*
HARVARD UNIVERSITY

undernourishment, morbidity, and possible mortality if he or she is unable to establish command over an adequate supply of food. For example, if a person does not have the means of buying enough food, he or she is not in a position to demand that food in the market. Insofar as the supply of food and ultimately the production of food commodities depend on market demand, this absence of demand will have a corresponding reflection in the lower supply and production of food. Superficially, while it may look as if there is not food enough for this person's needs, the cause of the problem may well rest in the shortfall of demand, and thus on the inadequacy of means to purchase, rather than on limitations of productive opportunities. In fact, in most economies in the world, the response of production and supply to prices and demands bears out the importance of the demand factor in the determination of output and supply.

In order to understand the nature of hunger, we have to examine the person's "entitlements" to food and related commodities. The starving person who does not have the means to command food is suffering from an entitlement failure, and the causal antecedents of this may lie in factors far away from food production as such. In each social and economic system, there are rules governing the respective rights that people have to exercise command over food and other necessities. In a private ownership economy, these rights are closely related to the ability to establish ownership over food and related goods. In order to understand the determination of entitlements, we have to examine the person's endowment position (e.g., what goods and services he or she can exchange for food), as well as exchange opportunities (e.g., at what rate the person can exchange what he or she owns for food). In the entitlement analysis of food deprivation and hunger, which I have tried to present elsewhere, especially in my book *Poverty and Famines*,[1] the crucial roles of both endowment and exchange entitlement relationships have been correspondingly investigated.

In going beyond market systems, other types of rights have to be considered in understanding entitlements over food that a person can enjoy. This can include claims against the state, e.g., the right to work (if such a right is acknowledged), the right to unemployment benefits (if there is such a social security arrangement), and so on. A person can establish command over food by making use of these different rights, and the entitlement analysis to hunger in general and famine in particular must take note of the institutional structure— covering legal and political as well as economic factors—that determine a person's entitlements, and through that, his or her ability to avoid starvation, undernourishment, and deprivation.

Two different aspects of the prevalence of hunger have to be clearly distinguished. There is, first, the problem of periodic famines, acute starvation, and mass mortality. Despite the enormous increase in productive abilities and national income in the world, there have persistently been frequent cases of severe famines across the globe. Famines in Ethiopia, Somalia, Sudan, Mozambique, the Sahel countries, Biafra, Bangladesh, and Kampuchea are just a few examples—many more can be given.

This problem of the persistence of famines has to be distinguished from that of the unrelieved continuation of regular hunger and undernourishment in a large part of the world. Most often, in fact, hunger does not take its toll in a dramatic manner at all, with millions dying in a visible way (as happens with famines). Instead, endemic hunger kills in a more concealed manner. People suffer from nutritional deficiency and from greater susceptibility to illness and disease. The insufficiency of food, along with the inadequacy of related commodities (such as health services, medical attention, clean water, etc.) enhances both morbidity and mortality. It all happens rather quietly without any clearly visible deaths from hunger. Indeed, so quiet can this process be that it is easy to overlook that such a terrible sequence of deprivation, debilitation, and decimation is taking place, covering—in different degrees—much of the population of the poorer countries in the world.

In understanding the causation of regular hunger and chronic undernourishment, on the one hand, and severe famines, on the other, the failures of entitlement have to be correspondingly examined. While regular hunger is largely a result of inadequate entitlements on a continuing basis, famines are the result of disastrous declines of entitlements, which typically occur rather suddenly. The landless laborer in a poor agricultural economy may be in a state of chronic shortage of command over food, and this will be reflected in persistent undernourishment of families in such occupation groups. This need not lead to dramatic starvation and immediate mortality, but the food deprivation makes the families more prone to disease, leading to a higher probability of death at a premature age. Seasonal unemployment, low wages, and other economic deprivations thus get reflected in health conditions and demographic facts. But all this may happen quite gently. In contrast, members of the same occupation group of landless laborers may die in large numbers if there is a dramatic decline in employment or if there is a sharp rise in food prices, ushering in a famine. In this case, the entitlement failure takes a much sharper and more severe form. Of course, both these processes may be observed in the same country at different periods of time. For example, in Bangladesh, the regular shortage of entitlements is causally associated with systematic undernourishment of the poorer rural population, without there being a famine all the time, but that population itself provided the bulk of the victims when the famine struck in 1974.[2] The economic processes involved in regular hunger and transient famine may well be quite distinct, but they both involve failure of entitlement over food and related commodities, in different forms.

Production and Entitlements

In understanding hunger, including famine, the focus of attention has to be on the dysfunctioning of the various links in the food chain covering production, distribution, and utilization, which are different aspects of entitlement relations. For example, famines may be caused by production failure, leading: (1.) to a direct decline of entitlements of those, such as peasants, whose means of survival depend on the food that they grow themselves; or (2.) to a sharp rise in the prices, thereby affecting the ability to command food on the part of those who have to buy food in the market.

But a famine can also occur without a decline in production, and indeed sometimes may take place in situations of peak food availability. In recent years, in fact, there have been many major famines without any decline in food output or availability per head.[3] The Bengal famine of 1943, the Ethiopian famines of 1973 and 1982, and the Bangladesh famine of 1974 are good examples. Indeed, during the Bangladesh famine of 1974, food output and availability per head in Bangladesh were higher than in any other period in the three preceding years or the two following ones. If one were to look only at output and availability as a basis for predicting famine, one would most certainly not have picked 1974 as the year of the famine; but that was indeed the year in which the famine struck.

In order to understand that famine—and indeed other famines—we have to look at the entitlements of the different occupation groups and how they moved over time. In the case of the Bangladesh famine of 1974, the victim groups were affected by: (1.) loss of agricultural employment as a result of floods during the planting and transplanting seasons; (2.) general inflationary pressure in the economy leading to a decline in real wages; and (3.) sharply rising food prices in the months following the flood, involving a good deal of speculative activities.[4]

The dissonance between the causal analysis of famines in terms of declines of food output and availability, on the one hand, and entitlement failures, on the other, does not lie in the fact that availability and entitlements are unrelated to each other. They are, of course, linked in many different ways. First, the output of food grown by some people themselves (for example, peasants) provides them with basic entitlements to food. Second, one of the major influences on the ability of anyone to purchase food is the price of food, and that price is influenced by the production and availability of food in the economy. Third, food production can also be a major source of employment, and any reduction in the production of food (due to, say, a drought or a flood) would reduce employment and wage income through the same process that leads to a decline, later on, of the output and availability of food. Fourth, if and when a famine develops, having a stock of food available in the public distribution system is clearly a major instrument in the hands of authorities to combat starvation. Public intervention can take either the form of direct distribution of food (in cooked or uncooked forms), or of adding to the supply of food in the market, thereby exerting a downward pull on the possibly rocketing prices.

It would indeed be amazing if food entitlements were entirely uninfluenced by food output, since the physical presence of food cannot but be an influence on the possibility of acquiring food through direct ownership or exchange. The dissonance does not arise from a denial of these important links. Indeed, the interconnection between the entitlement view and the availability view of food deprivation has to be considered along with their conflicts. The dissonance arises from the fact that the links do not establish a connection between availability and entitlement in such a way that the food commands of different sections of the population move up and down in proportion to the total availability of food in the economy. If food were to be distributed over the population on some egalitarian principles operated by some central authority, that assumption of proportional movements might well have been sensible. However, the actual command over food that different sections of the population can exercise depends on a set of legal and economic factors—including those governing ownership, production, and exchange—involving different parts of the food chain. It is thus quite possible for some groups (for example, a particular occupation group such as landless rural laborers, pastoralists, fishermen) to have a sharply reduced food entitlement, even when the overall availability of food in the economy is unaffected or increasing. For example, a decline in the price of animal products vis-à-vis cheaper calories in the form of food grains, or a reduction in the price of fish vis-à-vis the price of staple food, may adversely affect the pastoralists and fishermen, respectively. There have, in fact, been famines in which these changes in exchange rates have played a crucial part in the decimation of the respective occupation groups (for example, in the Bengal famines of 1943, the Ethiopian famine of 1973, and the Sahelian famines of the early 1970s). Changes in employment, wages, prices, and so forth can all take a major and decisive part in the initiation and intensification of famines.

The dissonance between the availability view and the entitlement view is particularly important to note in the context of economic policy. An undue reliance—often implicit—on the availability view has frequently been a contributory factor in the development of famine, by making the relevant authorities smug about the food situation. Policymaking requires a concern with each of the important links in the food chain, affecting command over food of vulnerable occupation groups.

Famine Relief, Food, and Cash

There are many policy implications related to the shift in the focus of analysis from production (and availability) to entitlements, in general, including the various links in the economic chains. One important question concerns the form of the relief that may be provided in helping famine victims. In African famines, relief has typically taken the form of distribution of free food in relief camps and distribution centers. While such relief has saved some lives, it has often been ineffective and inefficient, and the case for critically reexamining the appropriateness of this form of relief is not negligible.

We have to distinguish between two things that are achieved by food relief to famine victims: (1.) giving the destitute the ability to command food, and (2.) providing this relief in the actual form of food itself. Though these two aspects are integrated together in direct food relief for famine victims, they need not, in general, be thus combined. For example, cash relief can provide the ability to command food without directly giving any food to the victim. Indeed, cash relief can stimulate other parts of the food chain in terms of the response to increased demand as a result of the greater purchasing power of the famine victim.

A person's ability to command food can be seen as having two distinct elements, namely, his or her "pull" and the supplier's "response." If a person starves because of loss of employment and the absence of means of buying food, then that is a failure originating on the pull side. On the other hand, if the person's ability to command food collapses because of an absence of supply, or due to the "cornering" of the market by some manipulative traders, then that is a failure arising on the response side. In most famines—whether in Africa or Asia—the element of "pull failure" had tended to be the dominant one in the genesis of the collapse of entitlements of the famine victims. In this situation, creating purchasing power for the famine victim may be an obvious and immediate way of recovering some of the lost ground.

One of the big differences between the famine relief practiced in India and that typically used in Africa is the much greater reliance on cash distribution in India. While direct distribution of food is also used in India, a lot of the relief efforts in India, in fact, typically take the form of payment in cash wages for work. If the dispossessed find temporary employment and a cash wage, their ability to command food in the market is radically enhanced. Even if no food is brought to the famished by vehicles owned or requisitioned by the government, food does tend to move in response to the enhanced demands. The crucial issue, then, is to recreate the entitlement of those who have lost their means of support (e.g., as a result of loss of employment due, say, to a flood or a drought).

The famine relief efforts in India have been, on the whole, quite successful, and there has not been a famine in India since independence in 1947. In fact, the roots of this type of famine relief policy go well back into earlier periods—to the Famine Codes formulated in the 1880s—but often these reliefs were used only minimally or not at all in the pre-independence period. In fact, sometimes the Famine Codes were not even invoked (as in the Bengal famine of 1943). While the relief policies pursued in India in the post-independence period can be seen as extensions and refinements of policies that were worked out earlier, it is only in the post-independence period that famines have been effectively eliminated by the unvarying use of these policies, including the use of employment programs and wage disbursement (often in the form of cash payments).[5] Many threatened famines have been averted in different parts of India in different years using such relief schemes, e.g., in Bihar in 1967, in Maharashtra in 1973, in West Bengal in 1979, and in much of the north of India in 1987.

The use of cash disbursement has the advantage of quickness, which is particularly important in light of much-discussed delays in the relief system in the case of some African famines, such as those in Sudan,

Somalia, and Ethiopia. The provision of cash income leads to giving aid to the potential famine victims immediately. It also has the effect of pulling food into the famine-affected regions in response to the enhanced market demand resulting from cash disbursements.

The cash disbursement system also helps to prevent the widely observed phenomenon of what has been called "food countermovement," by which food moves out of the famine-stricken regions to more prosperous lands. This has occurred in many famines: for example, in the Irish famines of the 1840s, in the Ethiopian famine in Wollo in 1973, and in the Bangladesh famine of 1974. There is nothing terribly puzzling about this, since nonfamine regions (e.g., England in the case of Irish famines, Addis Ababa and Asmara in the case of the Wollo famine, and India in the case of the Bangladesh famine) often have greater purchasing power and greater pull in comparison with the famine-stricken regions. As a result, through the market mechanism, food can easily be attracted away from the famine regions to nonfamine areas.[6] When additional cash income is provided in the famine region, for example, through employment schemes, such "food countermovements" can be reduced or eliminated, and this itself may be very important.

There are, of course, also problems with providing relief in the form of cash income and employment schemes. Much will undoubtedly depend on the efficiency of the market structure in the economy in question. If the markets are so distorted that the expansion of demand will not lead to response, there may be no alternative to the government itself moving the food to the victims and directly arranging food distribution.

On the other hand, the administrative resources of the government are also restricted, and the record of famine relief based on large-scale movement of food grains by the government has not been particularly encouraging in many African countries. When the administrative structure is limited or inefficient, the case for using cash relief to regenerate entitlements and to create the pull for food movement may be a sensible policy. Indeed, in those African countries in which cash relief has been tried, such as in Botswana, the record of achievement seems to be very creditable.[7]

The system of cash relief also has some additional advantages. It regenerates the infrastructure of trade and transport in the economy through increased demand and more economic activities, and this can be of lasting benefit, especially since famines tend to disrupt these links in the food chain. Further, since the distribution to destitutes usually requires the setting up of "relief camps," the system of direct feeding or direct distribution of food can be very upsetting to normal family life, as well as to the pursuit of normal economic activities, particularly the continued cultivation of land. In contrast, providing relief in the form of subsidiary employment without making people move from their homes has the advantage of being less disruptive for work and living. While the decision on the balance between cash relief and food relief must depend on a careful and pragmatic assessment of the exact conditions in the country or region in question, it is worthwhile to consider the possibility of providing food to famine victims via the means of cash. When hunger and famines are seen not just in terms of availability of food, but primarily in terms of entitlement failures, it becomes natural to consider these other means of changing the operation of the dysfunctioning food chains. It is the broadening of the economic perspective related to this more general approach that may be seen as being the most important aspect of the proposed change of famine analysis.

Indeed, the form of famine relief is only one policy problem affected by the entitlement approach, and there are many other policy issues thoroughly dependent on the nature of the approach adopted.[8] Even in the context of assessing the form of famine relief (e.g., whether food or cash), there are many different policy instruments that can be considered, including the import of food from abroad, the introduction of rationing and control, taxing the incomes of nonvictims to give the famine-affected population a better competitive edge in the market, and so forth. These policy issues have to be examined severally and jointly in the formulation of public policy to deal with famine relief.

Famine Prevention and Productive Diversification

There are comparable policy implications for other aspects of antifamine policy, for instance, the prevention of famines (as opposed to giving relief to the famine victims). As was mentioned earlier, even in achieving this prevention, an alert system of distribution of cash wages and employment schemes may be important.

We have to distinguish clearly between food production as a source of income and entitlement and food production as a source of supply of the commodity food. The seriousness of the decline of African food production per head has often been seen as the main source of Africa's food problem. There is some truth in this diagnosis, but this is not only because food production is important as a source of supply of the vital commodity food, but also because food production provides employment, income, and entitlement to a vast section of the African rural population. The decline in food output has also led to an entitlement deterioration, and this deterioration has been brought about by several distinct influences acting together, including the loss of employment opportunities, reduction in incomes earned, and a general collapse of vital links in the food chain.

In enhancing the entitlements of the deprived and vulnerable groups of the population, expansion of food production can indeed, in many circumstances, have an important role. In fact, policy decisions regarding the relative balance of economic expansion in different sectors must depend on a careful calculation (taking into account the risks that have to be faced) of relative benefits that can be obtained from expansion in different fields.

A reasonable calculation of these relative benefits cannot be achieved, simply on the basis of plans to undo prevailing production trends—even if these trends are correctly seen as being responsible for the present state of affairs. For example, there are good grounds for giving priority to the expansion of food production in some parts of Africa, but that ground is not provided just by observing the historical fact that in these countries food output per head has declined, or that food imports have increased, or even that the decline of entitlements is directly linked to the production crisis. Policy decisions have to be based on assessments of the present circumstances and the anticipated future ones, and should not take the form of simple attempts to reverse the decline from past achievements.

In fact, insofar as the crisis of food production in Africa relates, at least partly, to climatic deterioration, that itself may be a good ground for considering other avenues of entitlement guarantee. The climatic deterioration must, of course, be encountered and halted. There is, in fact, a good chance that in the long run a more favorable economic and climatic environment for agricultural expansion in general and increased food production in particular may materialize, through programs of afforestation, irrigation, and so on. But economic policies should not be determined on the basis of imagining that such a change has already taken place. If it turns out that, given climatic uncertainties and the ecological problems, food production will remain very vulnerable to fluctuations in some parts of Africa for a long time to come, then it will be a mistake to rely too much on that one uncertain source of income and entitlements. This is an argument for putting greater emphasis on other types of production, including industrial expansion, from which more benefits may be derived, with greater certainty, in the present circumstances.

Entitlement analysis, focusing on different parts of the chain relations, points directly to the necessity of considering seriously the case for greater diversification of the productive structure of Africa. Alleged magic solutions of the problem of African hunger typified by such commonly invoked slogans as "put all the resources in agriculture!" (a favorite theme of "naturalists"), or "raise agricultural and food prices to boost production incentives!" (a favorite theme of many international institutions, including the World Bank), may deliver substantially less than they promise. There is, in fact, no escape from basing policy on a careful calculation of relative benefits—including the respective uncertainties—in choosing between different intersectoral balances.

Indeed, it is arguable that anti-hunger policies in Africa must give a very solid place to the expansion of nonfood production and the diversification of the economy. There is often a noticeable reluctance to consider the promise of industrialization for the future of Africa. Sometimes the reluctance arises from being unduly impressed by the favorable land-population ratio of most African countries compared with, say, Asia. But the choice between industry and agriculture has to be influenced by many considerations of costs and benefits, in addition to the availability of land. Climatic conditions themselves are a factor. The opportunity for economic growth that is provided by branching out into industries has been well demonstrated by the historical experience of many different countries in different parts of the world. Africa cannot ignore these opportunities. Also, the contributions that industrialization makes to skill formation and to the modernization of the economy and of the society may be important factors. The indirect influence of that technological transformation on agricultural productivity itself (including productivity in the food sector) cannot, by any means, be ignored.

As it happens, most of the successful agricultural economies in the world also happen to be industrialized, and this fact may not be a mere accident. Skill is as important an output for agricultural production as land, and the diversification of production seems to help the formation of skill. The favorable nature of the land-population ratio has, rightly, not been seen as a good ground for eschewing industrial production in Australia, Canada, or the United States. Dismissing that economic alternative for Africa on the ground of its high land-population ratio reflects, at the very least, some economic shortsightedness. While successful industrialization tends to be a long-run process, that process has to begin at some stage for it to bear fruit in the future. Also, the experiences of many developing countries have shown that in some branches of industry, rapid progress can be made by new entrants, yielding benefits without much delay.

All this is not to deny the importance of expanding food production in sub-Saharan Africa. Agricultural expansion in general, and that of food production in particular, will undoubtedly be one major instrument—but not the only one—in combatting hunger in Africa. Various strategic aspects of enhancing production in sub-Saharan Africa have received expert attention and scrutiny in recent years,[9] and there are many lessons to be learned from economic reasoning as well as from empirical observation of actual economic experiences. Given the number of people who derive their entitlements from food production in Africa, and given the limited speed at which this dependence can be reduced (though reduced it must be, in the long run), the importance of expanding food production in Africa, among the strategies to combat hunger on that continent, cannot be denied. It is really more a question of the balance between the different elements in anti-hunger policy for Africa, and it is in this context that the tendency to concentrate only on food production as a solution to the food problem in Africa has to be firmly resisted. The chain relations influencing food entitlements call for attention being paid to many other parts of the economic process. A narrowly focused concentration only on food production may well be wasteful and dangerous. The challenge posed by continuing hunger in Africa demands an approach that is adequately broad and balanced.

Concluding Remarks

In this paper I have tried to present briefly the entitlement approach to the problem of hunger (covering both regular undernourishment and transient famines), and I also have tried to examine the policy issues raised by the use of the entitlement approach. Only a few policy questions were explicitly discussed, but similar analyses can be presented in other areas of public policy, related to the general problem of hunger and the particular problem of famine.

Some of the areas of policy and action may take us well beyond purely economic reasoning. One of the particular fields of interest in the context of the problem of hunger is the division of food within the family, particularly between men and women, and especially between boys and girls.[10] There is some evidence from different parts of the world of systematic biases related to gender (e.g., greater undernourishment of women and particularly of female children). The sequence of chain relations does not end only with the purchase of food, since there is the further problem of division of food within the family, based on purchases made on behalf of the family as a whole. These problems of intra-household distribution linking the purchase of food in the market with the consumption of food by individuals in the family have to be investigated to get a better understanding of an important aspect of food deprivation found in some societies. Here again the notion of entitlement, related to the sense of perceived legitimacy, can be a major issue. Often, the greater share of men vis-à-vis women, or of boys vis-à-vis girls, reflects a traditional sense of what is legitimate and what is right. While these notions of legitimacy do not have the sanction of law enforced by the state, its grip on the actual distribution within the family may well be very strong indeed, enforced by convention and social pressure. There is considerable evidence of such traditional biases in the intra-household divisions of food.

There are many other issues in which the entitlement perspective is relevant. I shall not have the opportunity of pursuing them in this paper. The economic, political, and social factors involved in the food chains and their consequences call for careful scrutiny. What is needed is an adequately broad approach, instead of the narrow concentration on only production and availability.

There is no magic solution to the problem of food deprivation in the miserable world in which we live. But there is a unifying force—related to entitlement problems in food chains—that provides the basis for analyzing the diverse dysfunctionings that are present, and for seeking appropriate remedies. The overwhelming need is for relating policy and action to theory and understanding. This is an exacting task and its demands are large. But the rewards may be large, too. The survival and well-being of a substantial section of mankind may depend on it.

Notes

This paper draws on an earlier presentation by the author to a Brown University colloquium on "Perspectives on the History of Hunger."

1. Amartya Sen, *Poverty and Famines: An Essay on Entitlement and Deprivation* (Oxford: Oxford University Press, 1981).

2. M. Alamgir, *Famine in South Asia* (Boston: Oelgeschlager, 1980); A. Sen, *Poverty and Famines* (1981), Chapter 9; Martin Ravallion, *Markets and Famines* (Oxford: Clarendon Press, 1987).

3. See A. Sen, *Poverty and Famines* (1981), Chapters 6-10.

4. See particularly M. Ravallion, *Markets and Famines* (1987).

5. Amartya Sen, "Food, Economics and Entitlements," Elmhurst Lecture at Triennial Meeting of International Association of Agricultural Economists at Malaga, 1987, published in *Proceedings* and in *Lloyd's Bank Review*, 160 (1986); Jean Dréze, "Famine Prevention in India," WIDER Conference Paper, included in J. Dréze and A. Sen, eds., *Hunger: Economics and Policy*, to be published by Oxford University Press.

6. See A. Sen, *Poverty and Famines* (1981), chapters 7 and 10.

7. See J. Dréze and A. Sen, *Hunger: Economics and Policy* (forthcoming).

8. See A. Sen, "Food, Economics and Entitlements" (1986), and J. Dréze and A. Sen, *Hunger: Economics and Policy* (forthcoming).

9. See M. S. Swaminathan, *Sustainable Nutritional Security for Africa* (San Francisco: The Hunger Project, 1986). See also C. K. Eicher, *Transforming African Agriculture in Africa* (Cambridge: C.U.P., 1987); J. W. Mellor, C. Delgado, and M. J. Blackie, eds., *Accelerating Food Production in Sub-Saharan Africa* (Baltimore: Johns Hopkins University Press, 1987); F. Idachaba, "Policy Options for African Agriculture," WIDER Conference Paper (1987).

10. See Amartya Sen, *Resources, Values and Development* (Cambridge: Harvard University Press, 1984).

Summary of Discussions

Session III, Following Paper by Dr. Sen

DISCUSSANTS

Ignacio Narvaez Morales, Program Director, Global 2000 Inc., Khartoum, Sudan

Shlomo Reutlinger, Senior Economist, Advisor on Food Security, Office of Vice President for Africa Region, The World Bank, Washington, D.C.

Mogens Jul, Associate Professor Emeritus, Department of Food Preservation, The Royal Veterinary and Agricultural University, Frederiksberg, Denmark

Dr. Narvaez discerned a broadness of perspective in Dr. Sen's views, which he thought were nonetheless recognizably those of an economist. Confessing that he himself was by no means an authority on economics, Narvaez suggested that it was natural for an economist to consider hunger primarily an economic problem. This may hold true, he said, if the phenomenon is closely linked with related governmental policies formulated to enhance food production. The efforts to prevent hunger and famine, Narvaez asserted, should be exerted on a priority basis, with the first steps directed at achieving a sufficiency in food production. Once sufficiency is attained, economic measures can maintain or improve production. Until then, however, hunger is not primarily an economic matter.

Similarly, Narvaez considered the idea of entitlements per se to have limited or no significance in those rural areas that are periodically or dramatically affected by declines in food supply—as is the case in western Sudan, where most small farmers simply seek food security for themselves and their families, and seldom farm for commercial purposes.

In considering production and entitlements, Narvaez said, distribution is obviously the most important link in the food chain, even in situations where food is readily available. Narvaez cited the example of Sudan in 1985, when millions starved because of a failure in food distribution. However, when there is a drastic decline in food availability, distribution, and utilization, the economic implications are of lesser significance.

Amartya K. Sen; Robert Paarlberg; Ignacio Narvaez Morales; Shlomo Reutlinger; Mogens Jul

Narvaez felt that famine relief in the form of cash raises complex questions. Recalling the proverb that expresses the desirability of teaching a hungry person how to fish rather than simply giving him a fish to eat, Narvaez avowed a firm belief in the necessity of combining training programs with relief efforts. Although it has saved millions of lives during times of severe food shortage, Narvaez considered the P.L. 480 program was of doubtful long-term benefit, because he felt it did little to increase productivity or production of food. During times of food abundance or simple food deficiency, Narvaez conceded, productivity and production were indeed economic problems.

Dr. Reutlinger observed that Dr. Sen is not an ordinary economist. Most economists, lamented Reutlinger, are preoccupied with such matters as GNP growth rates, balance-of-payments, government budgets, trade policies, and so on, and there are few like Sen, whose basic preoccupation is with the economic and general welfare of people, and who examine these important economic decisions in terms of how they relate to the basic human right to food.

In Reutlinger's view, Sen's most important point lay not in emphasizing the fate of the hungry, but rather in admonishing us that, if this concern is not to remain hollow rhetoric, it must be translated into actions and public policies that make economic sense.

Reutlinger suggested that ethical notions of right and wrong lie at the heart of these public policies, and that the former must be altered in order to change the latter. Widely held beliefs are often irrational, he maintained. The same people who, on an intellectual level, can accept

Sen's recommendations, might hold very strong beliefs that lead them to support ineffective and sometimes counterproductive policies and actions. "Could it be," asked Reutlinger, "that ethical notions about what is right and wrong need to be constantly reevaluated and updated to embrace new economic realities? And, if not, do these outdated notions that served us well in the past, play perhaps into the hands of special interest groups when they are not updated and adjusted?"

Reutlinger illustrated his point with an example of a widely held belief holding that "to produce is to be virtuous; to produce food is considered exceptionally virtuous; and to receive entitlements or to augment the purchasing power in the market of those who earn less than what they need is sinful." Reutlinger believed that, in the real world, it is self-evident that some circumstances are beyond one's control, such as the population growth rate, the limited opportunities for economic advancement, and the consequences that, very often, people who work hard cannot produce enough for their own needs. Perhaps, Reutlinger reflected, it is time to reconsider this basic ethical perception—a perception that has created the alarming dilemma of the coexistence of food surpluses and persistent hunger. "On an intellectual level, I think most people see the problem correctly. People are hungry because they lack resources to produce enough food, or lack purchasing power to acquire food in the market."

Reutlinger pointed out that the victims of policies based on untenable ethical beliefs are not only in developing countries, however. Even in developed countries, an increasing number of farmers are in distress because food prices are on the decline; the supply keeps on growing faster than the demand. The policy failure, Reutlinger believed, is twofold. On the one hand, too little is done to enhance the ability of the hungry to produce or acquire more food. On the other hand, resources are squandered to pay all farmers to produce in excess of what markets will absorb, instead of providing direct adjustment assistance to those farmers who are unable to earn adequate incomes when food prices decline.

In Reutlinger's view, an imminent resolution of this problem is not in the offing, because public opinion favors subsidizing production and stocks over subsidizing consumption. The former is perceived as just, as a reward for hardworking citizens; the latter is perceived as relief and welfare for the unworthy. Moreover, Reutlinger noted, producer groups have political influence, whereas poor farmers and poor consumers, especially if they are far away in other countries, do not. It is not so much special interest groups that are to blame, he felt, but rather our pervasive, outdated beliefs of what is right and wrong. The result is that our country can spend $20 billion on producing surpluses which then go to waste, while simultaneously professing it difficult to find $2 billion to put purchasing power into the hands of those people and countries that experience desperate hunger problems.

Dr. Jul opened by expressing dismay over the prevailing paucity of political understanding, which he felt was even more serious than the absence of political will in addressing food issues. Jul said he was "flabbergasted," for example, by certain recent actions taken by European Economic Community (EEC) authorities, who displayed an outlandish adherence to bureaucratic decision-making procedures, which are based on information that moves at a glacial pace through administrative channels. A solution, Jul suggested, might be a greater concentration on the "science about people" that explains why politicians and farmers and family members act the way they do.

The suggestion that cash crops should be replaced with food crops, Jul felt, needs to be considered in a more mature fashion. In Denmark, for example, the acute starvation problems experienced in the later nineteenth century have since been solved by developing an export of cash crops. Other countries with little hope of self-sufficiency export non-agricultural commodities and import food. In today's global economy, all countries clearly do not need to develop food crops. In nations with unsuitable farmland, this would be unwise and counterproductive.

On the other hand, noted Jul, every country has some areas that are simply not as developed as others, just as every region has some nations that are less developed than others. Parts of the EEC, for example, are less developed than the rest, and it is taken for granted that the EEC nations collaborate in developing these areas. "I would like us all to look at the world in the same way," he said. "We are one world. We must collaborate in the development of those areas where there is a need."

Dr. Sunil Roy (self-confessed "environmental gadfly, necessarily unemployed") pointed out that most of the people who live on the fringes of permanent starvation are completely outside the economic field. For the majority of these people, Roy surmised, the Gross National Product means nothing. For them, the "Gross Nature Product" is the key to survival. Because the world has not begun to address the matter of the mere survival of these people, Roy observed baldly, millions of people are going to die before the end of this decade. The question Roy wished to pose to Sen concerned the quandary of "this vast number of people who are outside the whole field of economics—buying, selling; they don't have land; they don't have anything except nature to live on." How, asked Roy, do we tackle this basically moral and ethical issue?

Sen first reformulated Roy's question. Strictly speaking, this large group of people cannot be considered completely outside the economic field because, in order to survive, they have to somehow obtain food, either by growing it or by exchanging something else for it. On the other hand, many of them have absolutely no *assets* at all. As they have no land on which to grow food, the only thing they have to trade for food is their labor. Sen referred to these landless, rural laborers as a "most precarious group of people, the ideal famine victims." Their economic problem of how to obtain entitlements to food to meet their needs is different from that of other groups only in respect to degree, he felt. For people who have no assets other than their own labor power, Sen proposed that a solution to their entitlement problems rests, first, on the security of their employment and the kind of wages that they can expect to get, and second, on the existence of a social security network that should provide support if they cannot obtain employment. Roy's point, concluded Sen, concerns not so much whether poor people are outside the economic system, but rather, how the economic system should deal with their special problems.

James MacCracken sought advice from Sen about how private voluntary organizations might obtain wider public and governmental support for their efforts. The most urgent short-term need of such organizations, MacCracken suggested, was for money. In times of crisis, money supplies are rapidly depleted. During these periods over the last twenty-five years, MacCracken observed, P.L. 480 aid has proven a godsend. In contrast to Narvaez's earlier view of the P.L. 480 program, MacCracken perceived in it great potential and positive accomplishments through such strategies as "food-for-work"—even when times of crisis passed and the news media departed.

Sen distinguished three issues in MacCracken's inquiry. First, there was the question of shortages of resources generally (whether cash or food, either in-country or among would-be donors). Then there was the question of what private (rather than governmental) charitable organizations can do. Finally, there is the question of whether aid should be provided in the form of food or cash.

Sen addressed the last of these questions first, suggesting that there are some circumstances in developing countries where it is possible to create cash and disperse subsidies to people who would otherwise perish. This proposal, Sen recalled, "always sort of freezes people" on a basic level that has little to do with logic. Though such a practice may, admittedly, have an inflationary impact, Sen believed it is important to bear in mind that one is dealing with a situation where normal income has been lost, and cash aid might simply be considered as a temporary replacement for it. Demand is therefore not greatly multiplied. Moreover, Sen observed, "the only way you can get the market mechanism to move food from those who have it to those who don't have it, is to give those who don't have it more purchasing power than they have, so that they can attract food."

Such a strategy, Sen conceded, could only be employed by governments and not by private relief organizations. In Sen's view, the latter accomplish a great deal, though hardly on the magnitude demanded by the problems. Sen did not see this disparity as an argument against the efforts of private organizations, however. Until the necessary public efforts are mounted, private charitable work can prove to be of substantial benefit. "Just because it cannot solve all the problem, there is no reason for us to give up what can be achieved. There is no reason why the best should be the enemy of the good." On the whole, however, Sen claimed to be pessimistic about the possibility of solving the world's hunger problems through private means.

(Above) Hirshhorn Museum Auditorium, Smithsonian Institution, setting for Sessions I-IV of the colloquium
(Right) William J. Byron, S.J.
(Below) Yukio Yamada and Kiyoshi Ashida

(Above) Thomas R. Odhiambo and Norman Borlaug
(Right) Jessica T. Mathews, moderator for Sessions I and II
(Below) Nyle C. Brady and William J. Byron

(Above) Amartya K. Sen
(Below) Ignacio Narvaez and T. H. Roberts

(Top right) Robert Paarlberg,
moderator for Sessions III and IV
(Left) Mamdouth Gabr
(Bottom right) Ruben L. Villareal

(Left) Qing Wang
(Right) Setijati Sastrapradja

SESSION IV

Toward an Ethical Redistribution of Food and Agricultural Science

Surely, it is a major assault on our ethical sensitivities for hundreds of millions of people in developing countries to be so deficient in caloric intake that they cannot lead healthy lives, while at the same time there are other countries in which farmers are paid to reduce their food production, and in which uneconomically large stocks are held at high cost. Most people would agree that such a state of affairs is inconsistent with their basic philosophical position about equity.

Somewhat more subtle is the fact that in many food-deficit countries governments are markedly deficient in the scientific capacity for increasing basic food production, while the scientific base to generate food production is excessively large in other food-surplus countries. Thus, again we have an imbalance in world resources that is inconsistent with our moral and ethical standards.

Economists put in opaque terms what we intuitively know; namely, extra income means far more when a person is so poor as to be hungry than when income is at the American average, or even only slightly above subsistence. Of course, even at subsistence income, other items besides the cheapest calories are essential. Poor people know this, and they economize in consumption better than is generally understood. Thus, it is reasonable to equate inadequate food intake with the greatest assault on our ethical standards of fairness and justice.

Unfortunately, in the modern world we find that few steps have been taken to ensure adequate levels of food intake for all poor people. In general, the quantities of food given by the developed countries to the poor countries fall far short of the domestic deficit in meeting basic daily

John W. Mellor
Director
INTERNATIONAL FOOD
POLICY RESEARCH
INSTITUTE

needs, or in removing burdens and surpluses. Similarly, the level of foreign assistance to science made available to develop the agricultural potential of developing countries is far below what an efficient allocation of resources would suggest.

Why is there this continuing disjunction between our ethical principles concerning food and food production, and the reality of food needs in the developing world? I would like to suggest two sources for this inconsistency: technical and political.

On the technical side, it is indeed a difficult problem to transfer food that is in surplus in one area of the world to the poor in another area. There are physical problems in simply moving the food from place to place. There are also institutional problems involved in moving that food from towns and cities to the poor in rural areas. In many developing countries and particularly in Africa, the poorest people tend to live in the countryside, and it is in the countryside that food delivery systems tend to be weakest. Further, we do not know with precision what policies are needed to facilitate this movement.

On the political side, it is even unclear to what extent developed countries are committed to meeting the food and resource needs of poor countries. As Thomas Hobbes observed some three hundred years ago, sovereign governments tend to be far more motivated by narrow concerns of national self-interest than they are by broader humanitarian or ethical concerns. This means that in the area of food transfers, aid is often granted with the objective of advancing the political interests of donor countries. Such a practice of using food aid to advance donor interests other than humanitarian has the unfortunate side effect of overlooking those poor countries and peoples that are in the greatest need.

At the same time, it is striking the way in which the citizenry in the developed world fails to effectively pressure their governments to grant more aid. In the United States, public opinion polls over the last twenty years have consistently shown that some 55 percent of the general public favors economic assistance to other countries (*What Americans Think* 1987). Yet in the United States, such support for foreign assistance is far more passive than active. There are a number of reasons for this. On the one hand, many Americans are more concerned with fighting poverty at home than abroad. At the same time, a large majority of Americans feel that much foreign aid is wasted— either by the American bureaucracy or by recipient governments (*What Americans Think* 1987).

For these and similar reasons, there are no strong citizen lobbying groups for economic aid in the United States—or in most other developed countries. This is not to belittle the activities of many humanitarian groups, nor to deny the existence of lobbying efforts for food and other forms of aid in times of crisis, such as the recent famine in Africa. But these are clearly exceptional cases. In most instances, neither the rulers nor the ruled of most developed countries feel strongly motivated to push for increased amounts of economic assistance or food aid to the poor countries of the world. Note that the United States devoted 2.5 percent of its GNP to Marshall Plan Assistance to Europe after World War II, some 2.5 times the proportion of aid that the most openhanded government gives to foreign assistance today, and nearly ten times the proportion now given by the United States.

From the standpoint of the poor countries, there are other equally important political factors militating against closing the gap between our ethical concerns and the reality of ever-present food needs. Some poor countries fear that large food transfers will tend to depress future domestic agricultural production or bring demanding political dependency. The specific fear is that large amounts of cheap, imported food will depress domestic food prices, thereby either reducing the incentive for local farmers to produce more, or removing the premium for necessary government action with respect to agriculture.

Within the poor countries, there are also significant political problems involved in moving food aid and resources to those in greatest need. These problems relate to the very nature of many Third World governments. At worst, a number of Third World countries are authoritarian regimes—based on the rule of a narrow slice of military or ethnic elite. Anxious above all to stay in power, these elites tend to make sure that food and resources reach those whose disaffection they fear most: namely, the urban masses. Since they are geographically concentrated, urban masses can be quickly organized against the government (Bates 1983).

Rural masses, however, are more scattered and dispersed. As recent events in the Philippines, Haiti, and the Sudan suggest, widespread urban unrest is often the first step leading to the overthrow of unpopular Third World regimes. These regimes therefore tend to bestow more food and development resources upon town and city folk, even though the worst cases of privation are to be found in the countryside. But, even in more democratic Third World countries, the complex political processes of nation building may leave few resources for alleviating rural poverty. Thus, there are ample excuses for hunger existing amidst plenty.

Nonetheless, instead of dwelling upon these excuses, we must face an essential ethical issue: Why do we not devote ourselves more fully to finding ingenious means for removing these obstacles?

Food, Markets, and Instability

We live in a world in which food production fluctuates tremendously from one year to another and from one season to another. Modern market systems use the price mechanism to adjust to such fluctuations in production. Higher prices cause people to reduce their consumption of food, thereby keeping supply and demand in balance. If prices fluctuate sufficiently, they will induce some people to store food, and thereby transfer it from times of relative abundance to times of scarcity.

Yet, it must be recognized that there is an important ethical problem in relying on the market to equate the supply and demand of food. Because the poor spend such a high proportion of their income on food, the market places almost the entire burden of adjustment to changes in food supplies on the poor. Rising food prices redistribute income shares from the poor to the rich. For example, research in India (Mellor 1978) shows that food price changes translate quite directly into real income changes for the poor. Thus, while a 10 percent increase in foodgrain prices reduces the foodgrain consumption of the two poorest income deciles by 6 percent, it decreases the foodgrain consumption of the upper half of the tenth decile by only 0.2 percent. The absolute real expenditure on foodgrains is reduced ten times as much for the lowest two incomes deciles as for the upper half of the tenth decile.

Food price increases thus have a particularly deleterious impact on the poor, redistributing income away from them. Moreover, as food prices rise, the wealthier classes protect their food expenditures by reducing their consumption of those labor-intensive goods and services—such as transport and textiles—that provide employment for the poor. With fewer employment opportunities, the poor suffer a decline in their ability to procure food at any price.

To argue that the market acts unethically against the poor in the area of food price increases is not, of course, a general argument against using the market in allocation of resources. Indeed, the market helps to increase wealth so much that we want to be very careful that in interfering with the market for distributional reasons, we do not cause great inefficiencies in production—helping the poor now, but at their own later cost. We do indeed live in a complex world.

In view of these problems, it would seem desirable for governments to institute policies to reduce the worst effects of market fluctuation on the food consumption of the poor.

Two specific ways of doing so can be cited. In times of food scarcity, the rich countries can extend food aid to particular developing countries. This can be done either by bilateral programs from one government to another, or through multilateral agencies, such as the World Food Program. Alternatively, international institutions like the International Monetary Fund (IMF) can provide financing for food purchases, as it does now through its cereal import facility. While the practical impact of this cereal facility has been somewhat limited to date, in theory it is to provide poor countries with loans to buy food in times of domestic food shortages or high international prices (Ezekiel 1985; Adams 1983).

Of course, any mechanism for transferring food from rich countries to poor countries must be coupled with a mechanism within the poor country for transferring that food to poor people. Since we have suggested that in the short run market mechanisms do not work well for this purpose, some type of nonmarket mechanism may be needed. I will have more to say about this shortly.

Ethics and Food Transfers

It is clear that our ethical principles call for transferring resources to produce food as well as transferring food itself. Since the transfer of resources has largely to do with science, we will treat that in the section entitled "Science and Food." In this section, it seems useful to focus on the transfer of food itself.

Although food transfers in the form of food aid represent only a short-term palliative, such transfers are often unpopular. Many purists argue that it would be better to transfer money to poor people in the developing world, and then let them spend the money on food or other items as they see fit. I have little quarrel with this position. But let us note that we know the poor will spend largely on food and the food must be there as well as the money.

There are also those who maintain that food transfers have a negative impact on local food production (Schultz 1980). The argument here is that large amounts of food aid may reduce the local incentive of farmers to produce more food. Yet, empirical studies now suggest that the disincentive effects of food aid on local agriculture have been overemphasized. For instance, a review of twenty-one studies on the impact of food aid found only seven cases reporting "significant" disincentive effects on either prices or production (Maxwell and Singer 1979). In general, yesterday's large recipients of food aid are today's examples of agricultural success and commercial imports—with consequent large increases in per capita consumption.

On the more positive side, it is important to emphasize that food aid represents one of the best short-term policy measures for eliminating most of the low-end hunger in the world. According to the United States Department of Agriculture (USDA), some fourteen million tons of appropriately distributed food aid would bring consumption of the poorest peoples in most developing countries up to minimal nutritional standards. Fourteen million tons is some three-quarters of the total food aid given during the high period of food aid during the 1960s.

From an ethical point of view, it is painfully apparent that food aid should be made available to hungry people. Yet, there are those who question the effectiveness of such aid, on the grounds that only a small percentage of such aid—perhaps 10 to 20 percent—ever reaches those in need. Indeed, as noted above, many Americans openly question the ability and/or willingness of recipient governments to channel food aid to the neediest members of their populations.

Perhaps it is best to respond to these criticisms in two ways. First, from an ethical standpoint it is obviously better for poor people to receive a modest portion of a total food aid package than to receive none at all. Second, it would be useful if the more thoughtful critics of food aid would spend some time trying to devise solutions to these delivery problems. In countries like Bangladesh, it is doubtful if more than ten percent of food aid is improperly lost. How can that be reduced further and other countries brought up to a high standard of proper use?

In discussing the issue of how to transfer food to the poor, it is useful to divide the poor into the urban and rural poor. The rural poor tend to be far more numerous, both in absolute numbers and as a proportion of the total population.

In many situations, it is often possible to meet the food needs of the rural poor through food-for-work programs. These programs are designed to provide income-generating employment opportunities and to improve the infrastructure in rural areas. As the name implies, wages in food-for-work programs are usually paid in part or in full with food. In most cases, these wages are paid to laborers building new roads, drainage systems, and agricultural projects. Since improving the rural infrastructure is so essential to increasing domestic food production, it makes good sense to use food aid to fund food-for-work programs. However, if these programs are to have a long-term economic impact, it is necessary that they be coordinated with other policy measures and financing designed to make the infrastructure useful and productive over the long term. In general, food aid is not well coordinated with development strategy or finance.

Because of the employment provided, food-for-work programs tend to have a very favorable impact on the income of the poor. In Bangladesh, for instance, it has been estimated that the net income of participant households increased by 10 to 11 percent of their annual wage income, and by a much larger percentage during the season of their food-for-work employment (BIDS/IFPRI 1984).

While food-for-work programs may effectively provide income and employment for working members of the rural poor, there remains the problem of channeling resources to nonwage people in the countryside—specifically women, who head a high proportion of the poorest households. To some extent, poor women will be able to participate in properly constituted food-for-work projects. But, it may be necessary to have other supplementary programs to meet their specific needs. Programs catering to pregnant women and infants might be particularly desirable here.

Providing for the food needs of the urban poor may be something of a more long-term task. Some urban workers already enjoy substantially higher wage rates than their rural counterparts. However, many urban laborers suffer from the twin problems of unemployment or chronic underemployment. The problem then is to meet the more specific needs of these people while they search for work. This can often be done by instituting an urban food subsidy program.

In establishing food subsidy programs, it is important to realize that there is usually a critical tradeoff between expenditure on administration and targeting. In many cases, the administrative costs of targeting an urban food subsidy program outweigh the savings in leakages to the more well-to-do. For example, in Sri Lanka the food stamp program had no greater efficiency in reaching the poor than generalized subsidies.

This raises the problem of administrative resources. Many Third World countries are unable to meet the needs of all their poor simply because of administrative limitations. In the short run, these limitations can perhaps be overcome by the presence of foreign expatriates and experts, including private voluntary organizations. But, in the long run, this problems demands the rapid expansion of educational facilities at all levels. Developing countries must be able to produce their own cadres of trained, professional administrators. Those countries that have produced large numbers of trained administrators—e.g., India and Bangladesh—have also been best at meeting the needs of their urban and rural poor.

Long-Run Versus Short-Run Solutions

While food aid and food transfers are needed to meet the short-term needs of the poor, we must never lose track of the fact that in the long term we need to increase the earned incomes of the poor. And we need to increase food production, so that we have the food supplies needed to meet the ever-increasing demand of the poor.

In most situations, the best long-term way to increase food supplies and to raise the purchasing power of the poor is through a widespread pattern of technological change in agriculture. In general, technological change in agriculture can play two critical roles in the development process.

First, we need to recognize that food and employment represent two sides of the same coin. In the developing world low-income people typically spend 60 to 80 percent of their increments to income on food (Mellor 1978; Musgrove 1978). Thus, any strategy of development that leads to a rapid growth in the employment and income of the poor also leads to a large increase in the effective demand for food. If more food is not forthcoming, food prices will rise, the real cost of labor will increase, and investment will swing to more capital-intensive processes (Mellor 1976). Thus, any strategy of development that entails more employment for the poor will also require the wage goods—particularly food—to support such economic growth.

Second, technological change in agriculture has important employment and income linkages with the rural nonfarm economy. Technological change in agriculture raises the incomes of landowning farmers, who spend a large proportion of their new income on a wide range of nonagricultural goods and services. In Asia, for example, farmers typically spend 40 percent of their increments to income on locally produced nonagricultural goods and services such as textiles, transportation, and health services (Bell and Hazell 1980). The small enterprises that produce such goods tend to be far more labor intensive than any fertilizer factory or steel mill. They thus provide the rural poor with a whole spectrum of new nonagricultural employment opportunities. This increases the effective purchasing power of the poor at the same time that it provides for new rounds of growth in the economy at large.

A long-term strategy of technological change in agriculture, therefore, stimulates growth in the effective demand for goods produced in the nonfarm sector of the economy. Since that demand is for labor-intensive goods, it is very much equity oriented, in the sense of providing more food and job possibilities for low-income groups.

As noted above, technological change in agriculture can also do much to increase overall domestic food supplies. Yet, in many developing situations, even these increased supplies have trouble keeping pace with the ever-surging demand for food. Because of the high marginal propensity of the poor to spend on food, even those countries that have instituted widespread technological change in agriculture have been unable to keep pace with their rates of food demand growth. For example, the twenty-four developing countries with the fastest growth rates in basic food staples production between the periods 1961-65 and 1979-83 collectively increased their net imports of food staples by some 400 percent. These data show that the high rate of food demand growth that typically accompanies the development process is likely to exceed local production capabilities. The good news is that we can expand employment faster than a good record in agriculture can keep up with (with respect to supply of food). Faced with growing food shortages at home, many Third World countries in the high-growth stage of development are forced to rely on food imports.

It is here that the developed world has its most immediate role to play, namely, that of supplying food imports to the developing world to sustain high employment growth rates. Between 1961-65 and 1973-77, net food imports by the Third World increased nearly five-fold, from five to twenty-three million tons per year. Conservative linear projections of per capita production and consumption dynamics in the Third World suggest that the level of net food imports will reach eighty million tons by the year 2000 (Paulino 1986). More complex projections, taking the livestock sector more fully into account, suggest as much as 120 million tons of net food imports by the year 2000. This is the case for farmers in exporting countries to do well by doing good—by helping developing countries improve their agriculture and gain a growing market.

Science and Food

As suggested above, science is necessary to solve the future problems of food production in developing countries. Agriculture is a sector particularly subject to diminishing returns. As attempts are made to stimulate production, the inelastic supply of land causes the productivity of other inputs to gradually decline. Thus, each successive unit of output requires more inputs of other resources. The result is declining productivity, increasing costs, and accelerating poverty. The classical economist of the eighteenth century saw this as an insurmountable problem to which modern biological science would find the solution.

Fortunately, in recent years, scientific advances have helped make increasing land yields the engine of agricultural growth in the Third World. Between 1961 and 1980, output per hectare of major food crops in the developing world rose by 1.9 percent annually and accounted for more than 70 percent of total food production growth. During this period, increases in the harvested area averaged only 0.7 percent a year and contributed the other 30 percent of total production growth in the Third World (Paulino 1986). Although not yet realized on a substantial scale, yield increases are just as important in land-abundant Africa as the lowest cost means of raising abysmally low labor productivity.

In many instances, such encouraging advances in science are needed to offset contemporary high rates of Third World population growth. These rates of population growth demand far more of agriculture than was the case in the past. For example, between 1880 and 1920 Japan's foodgrain production growth rate was 1.6 percent. This equaled 160 percent of its population growth rate. Yet in more recent times, India achieved a rate of foodgrain production over 100 percent higher than Japan's rate. But, in India, the rate of foodgrain production was only 138 percent of the population growth rate (Mellor and Johnston 1984). With the burden of expanding foodgrain production becoming more and more dependent on obtaining higher yields, the demands on modern scientific technology are heavy indeed.

At this point, we should distinguish between using science and using prices to increase production. In a technical economic sense, prices cause production to increase, but at a decreasing productivity of resources. Scientific advances serve to increase the productivity of resources. Prices increase production, but at the expense of the welfare of the poor. On the other hand, science increases production with decreasing cost and therefore decreasing prices, to the benefit of the poor. There can be no disagreement about the ethical advantages of science as compared to prices in improving the food intake of the poor. Of course, there is a limit to the extent to which science can provide positive incentives to produce in the face of lower prices—but recent history shows how successful this can be.

Efforts to increase food production cannot be done with science alone. There also needs to be rapid growth in the infrastructure of rural development. This is essential in order to ensure that all segments of the rural population benefit from scientific advances. In general, the new scientific technology is scale neutral with respect to farm size. Science tends to benefit small farmers just as much as it benefits large farmers, provided, of course, that the necessary support services are in place. This means that government must take special steps to develop the support infrastructure—new roads, irrigation systems, and extension services—needed to deliver the benefits of science to all rural citizens.

Improved rural infrastructure that is so labor intensive (and therefore food intensive), may, in fact, represent the basic building block for accelerated agricultural development. The trained people needed to run new credit and fertilizer distribution programs will not move into the countryside until the basic infrastructure is improved. A common feature of the successes achieved in Japan and Taiwan is that each country established first-class research, good rural roads, comprehensive irrigation networks, and a broad education system. These are the basic elements for accelerated rural development.

Science and Ethics

Clearly, our ethical standards call for a substantial redistribution of science in the world. In 1980, public expenditure on agricultural research in developed countries was $4.7 billion, while for developing countries it was only $2.7 billion. There is a similar pattern of unequal distribution between developed and developing countries in the use of manpower in scientific research in agriculture. In 1980, developed countries employed scientist manpower equal to eighty-four thousand man years. Developing countries, on the other hand, employed scientist manpower equal to only sixty-four thousand man years (Judd, Boyce, and Evenson 1983). Since the term "scientist manpower" includes a greater number of people lacking postbaccalaureate training in the Third World, the qualitative differences between scientific research in the developed and the developing worlds are even greater than the numbers themselves suggest.

Clearly, the capacity to train scientists in food production science is also skewed toward the developed countries. Thus, there is a pressing need to give greater emphasis to training Third World scientists. As that stock of trained scientists is expanded, these people must be used to upgrade the educational facilities in their own countries. This means, first of all, developing good undergraduate institutions and then good graduate institutions.

We must remember that the problem is not just one of merely training scientists, but also of building the institutions in which they can work effectively. Building an adequate institutional base for education and training surely takes a minimum of ten to fifteen years. How this can be done on a modest budget, but with a long-term period and steady dedication is exemplified by the partnership in India of the Rockefeller Foundation and Indian research institutions. This partnership is one in which our honoree of today played an important role (Lele 1986).

In training Third World scientists, we need to be concerned about the proper mix between applied and pure research. Obviously, good applied science requires direct support from theoretical research. But deciding on the proper mix of resources that should be allocated between applied and pure research is a complex matter. It would, of course, be more efficient if pure research in the developed countries were more concerned with the needs and demands of applied research in the Third World.

Conclusion

In this presentation we have suggested that the people of the developed world seem not overly concerned with the ethics of food. On the one hand, one may argue that this is a misrepresentation of the situation, and that real technical and political problems impede us from realizing our ethical principles. But on the other hand, to date, our ethical inclinations have not caused us to put adequate resources toward finding solutions to these technical and political problems.

What must we do, if we wish to reform? The broad outlines are quite clear. In the short-run, we need to redistribute food from the surplus countries to the food-deficit countries of the world. Within the latter we must take steps to ensure that food reaches those who are in greatest need. It is not enough to move food from Iowa to Lagos or from Kansas to Manila; it also must be moved out into the rural hinterlands of developing countries. This means that developing countries must build administrative structures that are capable of efficiently moving food shipments to rural areas. This in turn will require more attention to the basic building blocks of rural development: roads, schools, and input delivery systems.

In the long run, there needs to be a major redistribution of scientific capacities from the developed to the developing world. Initially, this should take place through the judicious use of expatriate advisors to lay the foundations for scientific research and development in the Third World. But successful research can never take place in a vacuum. Developing countries must begin to pay more attention to the financial demands of scientific research. With respect to agriculture, recent work (IFPRI/ISNAR 1981) suggests that only twenty-five developing countries presently provide enough funding to support effective national research systems. If developing countries are to eventually produce their own cadre of trained scientists and specialists, they must devote more financial resources to building and maintaining effective research systems. Since they possess the resources, the developed countries can do much to encourage the Third World countries in this direction.

On the whole, it should be emphasized that we have both the resources and the expertise to redistribute food and science in the world. Although our past record at trying to effect such a redistribution is rather disappointing, several bright points do exist.

With respect to the short-run problem, it is important to recognize that we now have a major international agency—the World Food Program—designed to move large quantities of food to poor people. We should also note that the International Monetary Fund now has a cereal import facility, which is designed to finance movement of food to poor countries in times of need. There are also many public and private organizations—like UNICEF and the Red Cross—that do much to meet the temporary food needs of poor countries.

With respect to the long run, we should call attention to the efforts of people like Norman Borlaug, who was so instrumental in developing new high-yielding wheat varieties. We must also cite the efforts of the Rockefeller Foundation—beginning in Mexico in the early 1940s, later in India and continuing through to the 1980s—in its support for the creation of the Consultative Group on International Agricultural Research (CGIAR). CGIAR, which now includes thirteen research centers throughout the world, is having a major impact on the transfer of scientific capability from the developed to the developing countries (Baum 1986). And, of course, the record of Dr. M. S. Swaminathan in India is itself a bright spot.

Moreover, over the last twenty years, several developing countries—e.g., India, Kenya, and the People's Republic of China—have effectively pursued agricultural development programs that have significantly reduced the proportion of people living beneath the poverty line. Special programs for reaching the poor have also been implemented in countries like Bangladesh and Sri Lanka. Perhaps particularly striking is the unique combination of foreign assistance and developing country's effort in India, which used massive food aid to ameliorate short-run problems of food scarcity, even famine and inflationary pressures from growing employment, while the long-run problem of building the institution structure of agricultural growth was successfully erected. Again, M. S. Swaminathan played a key role in many stages and aspects of this long evolutionary process.

In conclusion, let us be encouraged by these and other situations where efforts are effectively underway to turn our ethical principles and standards into reality. With our present abundance of food and scientific resources, perhaps we can continue in the future to move toward a world in which hunger is no more and the worst of poverty has been banished.

References

I appreciate the assistance of several colleagues at the International Food Policy Research Institute, particularly Richard H. Adams, Jr., for his substantial work on this paper.

Adams, Jr., Richard H. 1983. "The Role of Research in Policy Development: The Creation of the IMF Cereal Import Facility." *World Development* 11, no. 7 (July).

Bates, Robert H. 1983. "Governments and Agricultural Markets in Africa." In *The Role of Markets in the World Food Economy*. Edited by D. Johnson and G. Schuh. Boulder, Colo.: Westview Press.

Baum, Warren C. 1986. *Partners Against Hunger*. Published for the Consultative Group on International Agricultural Research. Washington, D.C.: The World Bank.

Bell, C. L. G., and Peter Hazell. 1980. "Measuring the Indirect Effects of an Agricultural Investment Project on Its Surrounding Region." *American Journal of Agricultural Economics* 62, no. 1 (February).

BIDS (Bangladesh Institute of Development Studies, Dhaka)/IFPRI (International Food Policy Research Institute, Washington). 1984. *Results of the Household Survey of the Food-for-Work Program*. Prepared by Siddiqur Rahman Osmani (May).

Ezekiel, Hannan. 1985. "The IMF Cereal Import Financing Scheme." Report prepared for the Food and Agriculture Organization of the United Nations and the International Food Policy Research Institute. Washington, D.C. (August).

IFPRI (International Food Policy Research Institute)/ISNAR (International Service for National Agricultural Research). 1981. *Resource Allocations to National Agricultural Research: Trends in the 1970s*. Prepared by Peter Oram and Vishva Bindlish (November).

Judd, M. Ann, James Boyce, and Robert Evenson. 1983. "Investing in Agricultural Supply." Discussion Paper No. 442. Economic Growth Center, Yale University.

Lele, Uma. 1986. "Building Agricultural Research Capacity: India's Experience with the Rockefeller Foundation and Its Significance for Africa." DRD Discussion Paper No. 213. Washington, D.C.: The World Bank.

Maxwell, S. J., and H. W. Singer. 1979. "Food Aid to Developing Countries: A Survey." *World Development* 7, no. 3 (March).

Mellor, John W. 1976. *The New Economics of Growth*. Ithaca, N.Y.: Cornell University Press.

———. 1978. "Food Price Policy and Income Distribution in Low-Income Countries." *Economic Development and Cultural Change* 27, no. 1 (October).

Mellor, John W., and Bruce Johnston. 1984. "The World Food Equation: Interrelations among Development, Employment, and Food Consumption." *Journal of Economic Literature* 22 (June).

Musgrove, Philip. 1978. *Consumer Behavior in Latin America*. Washington, D.C.: The Brookings Institution.

Paulino, Leonardo A. 1986. *Food in the Third World: Past Trends and Projections to 2000*. Research Report No. 52. Washington, D.C.: International Food Policy Research Institute.

Schultz, Theodore. 1980. "Effects of the International Donor Community on Farm People." *American Journal of Agricultural Economics* 62, no. 5 (December).

USDA (United States Department of Agriculture). 1986. *World Food Needs and Availabilities, 1986-87*. Washington, D.C. (August).

What Americans Think: Views on Development and U.S.-Third World Relations. 1987. A Public Opinion Project of Interaction and the Overseas Development Council. Prepared by Christine Contee. Washington, D.C.

Summary of Discussions

Session IV, Following Paper by Dr. Mellor

DISCUSSANTS

Qing Wang, Director, Food Research Institute, Beijing, People's Republic of China

Setijati Sastrapradja, Director and Botanist, Centre for Research in Biotechnology, Indonesian Institute of Sciences, Bogor, Indonesia

Dr. Wang disclaimed any pretense of expertise in the fields of political science or economics, but observed that Mellor's paper raised issues of special significance to the political and economic concerns of her homeland. In a country with a population as large as that of China, it is particularly difficult, she said, to secure an adequate and nutritionally balanced food supply and to devise an equitable system of distribution. Wang noted that the problems of food distribution have proven especially intractable. For several decades, China has employed a central system of production planning and utilized a form of rationing for most food distribution. "We have paid a price in these procedures," Wang conceded. "We have sacrificed some personal freedoms for our achievements." Recently, however, China has introduced some elements of the market economy into its distribution system. A multitude of problems have been encountered in opening trade channels with the outside world, and in permitting agricultural producers free-market access to the cities.

It has been difficult, Wang claimed, to escape the necessity of providing development support in certain distributional and product areas. For example, substantial crop subsidies are provided to farmers, and huge sums are devoted to the transport of food from remote areas to the cities. Currently, Wang calculated, China allocates roughly $1 billion in subsidies to farmers—in effect, purchasing farm products at prices higher than those at which it sells them.

Meanwhile, Wang noted, a more laissez-faire economy leaves many older people at a disadvantage and permits some younger people to wield their new-found wealth in a manner disadvantageous to others.

John Mellor; Robert Paarlberg; Qing Wang; Setijati Sastrapradja

Wang expressed agreement with Mellor's contention that greater resources should be devoted to education.

Dr. Sastrapradja's remarks focused on the uneven global distribution of food and scientific capacities. Concerning short-term food redistribution strategies, Sastrapradja expressed agreement with Dr. De-Oliveria's earlier suggestion that charity should be avoided. "Charity," asserted Sastrapradja, "tends to make people lazy." Regarding longer-term efforts to develop local scientific capacities, Sastrapradja emphasized that commitments must be secured from developed countries to devote extended efforts to the kind of training that will be required to make a lasting impact. Training, moreover, must be relevant to local conditions. When studying in developed countries, she said, students must remain mindful of the kinds of problems to be faced in their own nations. Upon their return home, these students should not be sent immediately to remote areas. They will prove of more enduring value to their countries if they can maintain some communication with the worldwide scientific community and remain informed of recent developments in techniques and advancements in knowledge. Sastrapradja proposed that sharing of information and expertise between developing nations might, in the long run, prove even more beneficial than the training received by Third World scientists in the West.

Sastrapradja then discussed the links between food and education, two of the five basic human needs. Paraphrasing recent comments by the chairman of the Indonesian Institute of Sciences, Sastrapradja suggested that three basic economic groups could be characterized by their attitudes toward food. The first group's attitude might be characterized by the question, "Will I be able to eat tomorrow?" A second identifiable group has sufficient food security to be able to select the kind of food they wish. They would ask, "What shall I eat tomorrow?" A third group is the one in control of food systems and economic power. Their question would be, "Whom do I eat tomorrow?" These trichotomies in economic control over food, Sastrapradja emphasized, are intimately connected with education. As Sastrapradja said she stressed to a group of Ethiopian food authorities recently, "it is unfortunate to be uneducated, but it is doubly unfortuante if you are uneducated and you are poor."

Sastrapradja concluded with remarks on the relationship between ecological and technological security. Neither, she insisted, can be imported from outside. Ecological resources must be given careful consideration in developing appropriate indigenous technology. Water, for example, cited by Odhiambo as the key to the Green Revolution, varies widely in availability. In parts of Indonesia, three months of rain might be followed by nine months of drought.

Speaking from the floor, Alden Hickman, executive director of Heifer Project International, (and a Presbyterian minister), delivered a plea for "a theological-economic point of view." Though it may be better to teach a man to fish than to give him a fish to eat, Hickman observed, we must also consider who owns and who pollutes the fishing waters. Local political and economic conditions are too often ignored in devising foreign development strategies. At the heart of such efforts must be the conviction that we are indeed our brother's and sister's keeper. This conviction is conspicuously lacking, Hickman maintained, in much of the self-aggrandizing "aid and assistance" efforts of the great powers. "Unless we change the political systems that say, 'My country is more important than all these people without food,' we're not going to make changes."

Moderator Robert Paarlberg suggested that this dilemma might be summarized in the basic question of whether it is possible to have a just food system in an unjust society.

Patricia Kutzner directed her comments at the issue of control over research priorities. "An enormous amount of investment is going into science by many of the corporations, who expect to patent the results and thereby, obviously, reap the profits from their investments." This, she admitted, is "logical, very logical." But corporate research efforts are generally not directed at the needs of small subsistence farmers. The research of international assistance organizations may be geared more toward these needs, but Kutzner asserted that little attention has been given by the scientific community to the question of how to distribute the fruits of technological research to those most in need of it. Given the necessary funding, one can always direct research toward what one feels is important, Kutzner observed, but the pertinent questions are "who makes those funding decisions, and on what basis, and for whose benefit?"

Sunil Roy responded by cautioning against what he felt was a very real danger of complacency. Though India is often cited as a great success story, for example, Roy insisted that we must not forget that there remains an enormous gap in providing for the nutritional needs of the population. Roy pointed out that there are about 350 million people in India who get less than 75 percent of their minimum calorific needs. Of that, he said, some 12 percent get less than 50 percent of their minimum calorific needs. The achievements of Swaminathan and his colleagues notwithstanding, terrible hunger problems remain, said Roy. Shortages of food in this regard are relative. As Mahatma Gandhi observed, "There's enough for everybody's need, but not for everybody's greed." From the standpoint of India, a plain need exists for outside analysis and understanding of the food situation, but control must remain in Indian hands. "Anyone who comes with readymade solutions and advice must first understand from us what our questions are and help us to articulate them better, and then help us jointly to search for solutions together," declared Roy. "He alone is a friend who helps us to think about our problems on our own."

John Mellor brought the session to an end with "a few fairly simple statements about what is innately an extraordinarily complex issue." First, he noted, as long as the poor remain relatively poor, food will be a major expenditure for them. From that standpoint, food will therefore continue to be a rather special commodity. Second, because of the constraints on the availability of land, agriculture will remain peculiarly dependent upon science for increasing productivity. It is relatively easier to increase production of most other goods and services without the contributions of science than is the case with agriculture. Conversely, agriculture will be particularly susceptible to rising costs if science does not come to the rescue. Compared to science, Mellor averred, price increases are a relatively inefficient means of bringing about increased production.

"So if we are concerned about poverty issues," Mellor continued, "we particularly want to look at the possibilities of bringing science to our assistance." Granted, applications of science to agriculture bring with them a host of problems, but these should not deter us from seeking scientific solutions in situations where they may be most desperately required. Hazarding what he termed a "gross simplication," Mellor suggested that "scientific advances in agriculture give us percentage increases in production, not absolute increases." The distinction, said Mellor, was that a scientific breakthrough in areas where yields are already very high will give about the same percentage increase as a scientific breakthrough, or a certain amount of expenditure on sciences, in areas where they are low. The absolute increase would accordingly be much larger in the area already relatively well off. Admitting that there may be exceptions, Mellor proposed that science has a tendency to widen productivity differences among regions. Combined with the powerful multiplier influences of agricultural growth, this will greatly exacerbate regional disparities in aggregate income. Science can therefore prove a double-edged sword, Mellor cautioned, and care must be taken to devote extra assistance to deal with the problems of the very poor in those parts of the world where the most productive agricultural areas also have a large landless labor class. Given the underlying tendency of science to widen regional disparities in income, there is the inclination in some nations to devote most of their scientific resources to the poorest areas, and very little to the better areas, presuming that the two will then move more-or-less together. Such a policy, Mellor warned, will prove greatly detrimental to overall growth.

Mellor then cited a number of studies conducted by the Internatonal Food Policy Research Institute that indicated that "if you want to help the poorest people in the more backward areas, you do more for them by obtaining the much greater growth you can get in the more forward areas, and having the very unpleasant process of population migration." Mellor conceded that migration from rural to urban areas did indeed produce an "unpleasant" distribution of age and skills in populations of both regions. Still, Mellor maintained, it may be more pleasant than the alternatives. All income distribution problems cannot, after all, be solved by scientists. When the scientists get through, quipped Mellor, there ought to be a little bit left for sociologists and economists to do.

SESSION V

Summary of Discussions

Dr. Swaminathan agreed to serve as moderator for the final session of the colloquium, during which no papers were presented. According to the colloquium prospectus, discussions during this session were "designed to draw conclusions and establish an agenda for future international action on food issues."

Swaminathan began by outlining three broad areas he had discerned being raised in the colloquium up to that point.

First, there was the question of how to deal with hunger problems associated with relatively short-term prospects, but not directly concerned with emergency relief. There seemed little dispute that enough food exists in the world to provide for the nutritional needs of all, so in Swaminathan's view, the challenge lies in how to distribute it to those who need it while avoiding damage to existing agriculture.

Second was the question of how to build up national capabilities in the Third World. How, he asked, given current trends in the developed world, is this to be accomplished? A number of issues raised previously are involved in this question, he thought, including the transfer of scientific resources, the development of national infrastructures of various kinds, the devising of national policies for effective food distribution, and the development of self-reliance in food production.

The third major area concerned the long-term sustainability of the above processes, on an ecologically sound and economically viable basis. Admitting that there were doubtless more areas of great concern that might be addressed, Swaminathan proposed that these three form the basis for discussions in the last session.

Led by M. S. Swaminathan

Focusing for a moment on the third of these areas, Patricia Kutzner interjected that social sustainability should also be considered, particularly so that the poor are assured some direct participation in the processes that have such a great impact on their lives. Ignacio Narvaez later suggested that the political point of view should be kept in mind as well. Even with the best of implementation policies, and programs utilizing technological advances, nothing will happen, he maintained, if there is no political stability.

Encountering no opposition to his proposed discussion outline, Swaminathan proceeded to the first topic—that of short-term, immediately feasible approaches to the problems of hunger, malnutrition, and health. He urged that participants keep in mind the discussion objective of developing some practical recommendations for future action. Perhaps a mental balance sheet can be drawn up, he suggested, so that some assessment might be made of the relative strengths and weaknesses of current efforts aimed at ending endemic hunger, chronic undernutrition, and poverty. From the floor, it was suggested by a representative of a private voluntary organization that there is a pervasive lack of knowledge among many involved in these efforts about what their colleagues in other fields are doing. There is a particularly alarming absence of a shared body of knowledge between the scientific and voluntary relief communities, he said, leading to the danger of duplication, rather than reinforcement, of efforts.

In contrast to Dr. Narvaez's earlier stated views, Nevin Scrimshaw believed that political implications should be considered only after one

The International Center, in the Smithsonian Institution's S. Dillon Ripley Center, setting for Session V

has defined the biological and social nature of the problem. "The basic biological and social problems that we're trying to correct relate first to the whole issue of chronic deficiencies of protein and energy." Scrimshaw referred to populations "in the hundreds of millions" forced to adapt to low-energy intakes by reduced physical activity, with attendant social and economic consequences. Children, for instance, have to adapt to reduced energy intakes with deleterious effects on their learning behavior and cognitive development. The long-term consequences of this adaptation, said Scrimshaw, are a reduced physical growth and development and a lower adult stature. "Any idea that adaptation to energy deficiency is without social, economic, and personal cost," Scrimshaw protested, "is nonsense." Ensuring adequate protein in diets remains very much an issue, he said, and we continue to see the results of protein deficiency in children today. Iron deficiency too remains widespread, with consequences that include not only increased susceptibility to infection, but decreased work capacity and productivity, as well as damage to cognitive development and learning capacities. Iodine-deficiency disorders, such as goiter, are heartbreakingly common in parts of the world, as are vitamin A deficiencies.

Sunil Roy broached the subject of "food-for-work" versus cash as a form of interim relief. Cash, Roy claimed, often cannot be used to buy food and is not really a solution. Ecological problems connected with this include deforestation by poor people seeking to sell wood in order to obtain food. Such disenfranchised poor are without even the meager resources to make productive use of what Roy referred to as "wasted land."

William Mashler expressed surprise at not having heard any mention of the "paramount" issues of education in discussions on Africa thus far. "Hunger and lack of education go hand-in-hand," Mashler insisted. The areas of the world suffering most from malnutrition, he said, are those where education is least prevalent. Education therefore must be a cornerstone in any programs that presume to have a permanent effect on hunger. Though education and hunger have often been discussed in isolation, no aid program to date has looked at the two issues in their totality as being complementary. This applies equally, Mashler suggested, to the nutrition dilemmas discussed by Scrimshaw. If the health of children is to be improved, asserted Mashler, then something must be done about the education of their mothers. A high-quality maize germ plasm might be produced for replication, but the maize protein will never reach children's stomachs if there are no producers with an adequate education to make use of it. In the area of medicine, there is clear evidence of a direct relationship between diet, diarrhea, and human reproduction.

Mashler also wished to second Father Byron's statements about the importance of ethics in the global food system today. Having worked in both the international and national systems, Mashler reported he witnessed "a horrendous deterioration in the ethical behavior of the human race." Ethics, he warned, need to be reawakened in many areas: in the family, in the schools, and in the universities. The leadership in our countries, political and otherwise, must determine what current ethical standards are and what they ought to be, and then be able to align the current with the desired standards. "If they don't, I see a grim future for our world."

Eugene Donefer reflected on the novelty of a colloquium that attempted to discuss ethics in combination with science and food. He perceived the repeated references to the concept of "the right to food" as a noteworthy feature of the colloquium.

Donefer emphasized that hunger is a worldwide problem and is not restricted to Africa and Asia. It also remains a serious problem in the Americas, including Canada and the United States.

Norman Borlaug remarked that the psychology of change is an important subject that has been largely overlooked. The peasant farmer, for example, has frequently been accused of being a traditionalist and resistant to change. Borlaug insisted that this is quite incorrect, and one must be aware of that when attempting to apply the results of the laboratory to the farmer's fields. When on-farm tests were being conducted in Pakistan and India during the mid-1960s, heated arguments developed between biologists and economists about the levels of nitrogen and phosphor. Borlaug and his colleagues, intent on showing what kinds of yields were possible, insisted on using levels of these chemicals far in excess of what should or could be normally used from the standpoint of environment and economy. In due course, these demonstrations ignited what Borlaug referred to as a "grass-roots fire." Great enthusiasm was generated among small farmers who suddenly discovered that their traditional methods were wrong. Instead of producing one ton per hectare, they could produce five or six tons on the same piece of land. Only after small farmers had been convinced of what was possible, Borlaug said, could political leaders be persuaded to change economic policies. Credits were then made available to small farmers, so that they might buy seeds and equipment at the time of planting and pay off their loans at harvest time. This ultimately enabled farmers to sell commodities at a price close to that of the international market. By announcing at planting time that farmers would be guaranteed purchase of their products at market prices if their goods could not be sold in the private sector, the government triggered a vast growth in the farm economy.

Borlaug suggested that the only force that can induce government officials to make the appropriate economic moves at the correct times is the pressure of huge numbers of the common people. In India, 200,000 to a million small farmers have been exerting that pressure, though it remains to be seen how successful they will be. Whatever the outcome, we should not underestimate the factors involved in the psychology of change.

Borlaug also expressed agreement with Mashler's comments on education. From his observation of sub-Saharan Africa, Borlaug concluded that education, particularly at the primary level, is one of the major bottlenecks to progress. In the back country, there appeared to be no formal education at all. "The whole infrastructure is miserable," he lamented. Borlaug did discern "real opportunity for a breakthrough in one or two countries." Cautioning against the tendency common in foreign technical assistance programs to try to correct all ills of all countries at once, Borlaug suggested that recognition be given to the heterogeneity of Africa's climate, soil, and cultural and ecological conditions. By selecting two or three places for on-farm demonstrations, using appropriate technology already developed at government experiment stations, breakthroughs in production can be achieved. These in turn will place great pressure on government leaders to implement sensible policies, thereby leading to significant changes. Available technology simply has to be used in a coordinated fashion, Borlaug insisted.

Emil Javier returned to the question of how ethics can be incorporated in the development goals of a given country. In many places, he noted, there is not even a consensus on what the national development goals are. Even assuming that a consensus can be reached, Javier believed there would remain a great need for the "development of an analytical capacity to analyze what the options are within the available resources and the culture." Javier feared that many countries simply don't have the capacity to include in their development policies an educated assessment of economics, social and political conditions, and scientific factors in order to assess what their priorities ought to be. A developing country's first step, therefore, should be the development of a capacity for analyzing its own needs.

Referring to Kutzner's earlier comments, Javier affirmed that one must include the people one is trying to assist in the process of policy development. Even in the poorer countries, Javier suggested, there is some capacity for scientific research. At the micro level, the so-called "farming systems perspective" or "back-to-farm approaches" must be considered when establishing research priorities.

Jim Phippard concurred with Javier's assessment of the need for policy analysis capability in countries concerned with food aid. In the development of the agricultural sector, said Phippard, "all the good will in the world won't work if policies aren't set right." Phippard wished to emphasize, however, that short-term and long-term food aid policies should be considered together. At the moment, for instance, certain of the multilateral development banks and the bilateral donors are undertaking structural adjustment programs, particularly in Africa. It is common knowledge, he said, that the burden of these adjustment programs, and the austerity measures they produce in debtor nations, fall most heavily on the poor. Of course, some consideration should be given to the employment ramifications of debt adjustment, Phippard cautioned. In some countries, bloated public sectors have been reduced by these programs. But food-aid policies should also target those most likely to be put out of work by such programs. Food-for-work is one strategy for dealing with these situations, Phippard suggested. Income-generating projects, especially, should be developed for people who will be hit by austerity programs. By identifying, for instance, a particular factory that would be likely to close down in the event of an inefficient public sector organization, one might devise for those employees certain income-generating projects funded from the sale of food-aid commodities. By such means, food aid would not flow directly to the needy, but would allow the needy to derive income indirectly for purchasing food. Phippard stressed the benefits of sustainability in such income-generating programs.

Patricia Kutzner pointed out that, when talking about family nutrition, "it's the women in the household that you have to empower, primarily, with whatever food-aid policies you pursue." This will involve a different meaning and require a different strategy for each culture, she noted. Nevertheless, in almost all cultures, women are the grass-roots experts in food distribution, as they are inevitably the most knowledgable about the food system, and they are the ones who must deal with it. Kutzner drew attention to new research developed through the United

Nations Women's Decade, which might assist in reaching this grassroots expertise. Most countries now have strong women's organizations that might be used as channels, she indicated. More information on these is available through the UN Development Fund for Women, UNDP.

"We will not succeed in any kind of food-aid program in the nutritional sense that we're concerned about unless the local women are involved in the planning and administration of that program," Kutzner maintained. She warned that food-for-work projects, however, usually add burdens to the women of the household, "who have to go out and work on those roads and those dams, while they try to feed and heat and gather fuel and bear children, and so on."

Mogens Jul, who characterized himself as "a very pragmatic sort of a guy," professed to abide by one principle: "It doesn't really matter to get the right thing said; it's to get the right thing done." Jul felt that it didn't much matter whether one spoke of the long or the short term. The real problem is to convey one's ideas to policy formulators and decision makers. Among EEC decision makers today, for example, the prevailing attitude is that "food aid is bad." "Since we have mountains of food to give away, it's a rather unfortunate attitude, but it must be addressed," said Jul. Decision makers in the Third World, Jul noted, often are motivated by the vague notion that they should imitate the Western world. It will take the influence of men such as Swaminathan or Borlaug to puncture these fallacies.

Barbara Huddleston had two observations to make. First, she suggested that one way to "restimulate the ethical conscience" would be to start a campaign to ratify the United Nations charter on human rights. Not many realize, she said, that the United States has not yet ratified this charter. A campaign of this sort, she felt, would be a way to reawaken the debate in the United States on the right issues: the importance of preserving the vision of one world, of eliminating We-Other opposition, and of embracing the point that we are our brothers' keeper in the joint sense of community. By not signing the UN human rights charter, the U.S. was implicitly refusing to subscribe to the principle of the right to food. Huddleston felt that this greatly weakened the U.S. position in the world community.

On the question of assessing requirements for food aid, with which Huddleston's work at the Food and Agriculture Organization (FAO) is directly concerned, she suggested that emergency supplies of food are "really not such a big problem" in the short term. When a need can be demonstrated in an emergency, suppliers of food will usually come forward. The biggest problem, Huddleston felt, involves logistics and the problem of moving the food to the people who need it. This is particularly serious in situations where there is civil war or a military disturbance of some kind.

Huddleston conceded that the problem of supplying food aid while other development processes are taking place is a much more difficult one. This, she said, is the area of great challenge for food aid. Referring to Mellor's earlier remarks, Huddleston noted that the food alone can only account for perhaps 15 to 40 percent of the total resources required for development activity. This food aid is usually not coordinated with other kinds of activities. Subscribing wholeheartedly to calls for more interdisciplinary action, Huddleston drew attention to the folly of, for example, trying to stimulate production in a region where, at the same time, the landscape is being deforested. "The one hand takes away what the other hand has given."

Donald Duvick observed that little had been said about the role of private industry in the problems discussed. Private industry, he pointed out, does specialize in delivery systems, and it therefore deals closely with local populations. It operates both on a large scale and on a very small scale. It's concerned with indigenous local companies and with mammoth transnational companies. At times, he noted, it even practices good ethics—though at times it does not.

Joseph Hulse pursued this thread. Particularly since General Foods was underwriting the colloquium costs, it was indeed surprising, he asserted, that virtually nothing had been said about what happens to food after it leaves the farm. We don't eat raw grain, after all, nor do we eat live animals. We process them. The growth of agriculture, especially between the two world wars, may have led to the decrease in the number of people on U.S. farms from about one-half of the national population to only 4 percent, but many of those departing farmers remained at work somewhere in the food chain. A great many more people than 4 percent of the population of the U.S. today are concerned with feeding the country, Hulse noted. Agribusiness—the processing of agricultural products—has, in fact, accounted for the economic growth of almost every nation, with the exception of Singapore. However, Hulse pointed out that there are very few food technologies being used today that weren't known in ancient Rome. Essentially, the food industry has done nothing more than mechanize and develop on a larger scale what was originally a kitchen craft. The modern flour mill is exactly the same in principle as those that were known in ancient Egypt at least six thousand to seven thousand years ago. Today's most common processed foods are those that have long been known. Even the lowly hamburger is mentioned by Apicius, in his "De Re Coquinaria," as having been sold on the streets of fourth-century Rome.

Hulse's point was that employment in the food industry is intimately connected with agriculture per se. Though the question of food-for-work had been repeatedly raised, "I'm talking about work-for-food," he said. An important component of this sector, Hulse maintained, consists of small food processors and vendors. Often governments attempt to drive this sector off, but Hulse believes that it is much wiser to follow Singapore's example of making the practices of these small businesses more efficient and hygienic. One doesn't need a Ph.D. in food science to run a bakery or a flour mill, or to do most simple food processing, Hulse insisted. A great deal more effort could therefore be given to getting the landless poor into the food processing business. Though it is not complicated, it does require management skill—and that is an area too often neglected by university food science departments. How small food businesses should manage money, material, and, particularly, human resources, are important problems that have not been adequately addressed by food science. These skills are certainly more urgently needed than is the invention of a new groundnut sheller in India, or yet another on-farm grain storage invention for Africa. Though it is doubtful that many more inventions are needed, the emphasis in our teaching nonetheless remains on inventiveness and innovation, rather than on usefulness. At a recent meeting of the International Union of Food Science and Technology, Hulse noted, a resolution was passed in the executive session that called for greater attention to be paid over the next several years to the development of small-scale food industries, and the training of people to work in them.

L. K. Bharadwaj's comments concerned the importance of education in development projects. For example, in the University of Wisconsin's Pushka Project (devoted to fighting desertification in nothern India), "we are trying to use education and development as critical inputs for employment generation." As part of this effort, people are being taught production, marketing, and management skills, "so that they become self-sufficient with very low levels of technology and very low levels of income inputs."

Bharadwaj also wished to draw attention to an idea put forward in 1976 by Swaminathan, who proposed that the Indian government establish a "rural reserve corps" of agricultural and marketing experts. If this idea could be revived, suggested Bharadwaj, it "could be the vehicle for inculcating the scientific and ethical attitudes that we talked about, and at the same time it would become an attempt to reverse the brain drain from rural to urban areas that Mahatma Gandhi talked about." It could also be a vehicle for involving people in development projects, he indicated.

Bharadwaj too expressed a hope that more projects would be developed that aim at income generation, "so that people can buy food, rather than simply look for piecemeal solutions, either in cash given to buy temporary relief, or in work provided where there's no need for it."

Swaminathan at that point suggested that the group turn to the second discussion issue proposed by him at the beginning of the session.

First, however, he attempted a summary of discussions up to that point, "so that we can see if we have a consensus as to what we want to recommend." As several speakers noted, he said, the first problem is of worldwide scope, encompassing developed and developing nations alike—though the nature of the problem might vary. Secondly, Swaminathan noted that all seemed to accept the notion that the right to food should be an integral part of national food policies. This right would include both the physical and economic access to food. In this sense, the term "food" should be understood to include all the elements of nutrition that are needed for a balanced diet.

Given this overall understanding of the problem, what then are the steps that are needed to address the issue? Swaminathan observed that many speakers emphasized the need to understand the problem at the micro as well as the macro level, in order to find a meaningful solution. At the micro level, he cited both the call for increased anticipatory analyses of issues and Borlaug's stress on the importance of understanding the psychology of change. With respect to shorter-term strategies, Swaminathan referred to food-for-work, as well as to what Hulse termed "work-for-food." Employment generation strategies, on the other hand, involved both short-term provisions of immediate work and long-term mechanisms for permanent employment. Swaminathan stressed the crucial role of education in these programs. Health problems were mentioned, he noted, as were the problems involved in commodity delivery systems, food processing, food storage, and the availability of a rural resource corps of professionals who could help in implementing food strategies. In discussing implementation mechanisms, Swaminathan said, principal emphasis had been placed on the capacity of governments for policy analysis. Finally, summarized Swaminathan, the role of the public sector, of private industry, and of people's organizations had been discussed. Suggestions for different methods of employment generation had been put forward, including the training of managers for small-scale industry.

Unfortunately, Swaminathan concluded, "we have probably gone around the problem once again." Starting from the premises that hunger is a global problem and that there is a universal right to food, any meaningful discussions must perforce examine possibilities for solutions at the local level. In Swaminathan's view the above-mentioned mechanisms of education, health, and delivery systems will probably have to be related in some manner to the national policies and strategy analysis capabilities. Ultimately, he maintained, "a national problem essentially has to find national solutions."

Nevin Scrimshaw hastened to add that "the whole problem of hunger and malnutrition is multisectorial." Though there is a tendency to relate it solely to the agricultural or economic sectors, the problem must also be related to the education and health sectors. "Part of it is going to be approached through the health sector—through primary health care, through child survival strategies, through programs of immunization, weighing of children, and the rest."

John Mellor apologized for suggesting that the list of items summarized by Swaminathan were "mostly points for lecturing developing countries about what they should be doing." Though he felt it was not inappropriate for a group as international as this one to take this attitude, Mellor pointed out that very little lecturing had been done to developed countries about what they should be doing.

Mellor suggested that a broader view of the situation should be considered. First of all, he said, let us face the fact that there are a tremendously large number of people in the world with alarming food deficiencies. Experience has demonstrated how very long it takes purely domestic efforts to solve such problems in Third World countries. It is easy for those in developed countries to assert that there should be a redistribution of wealth in poor countries, but few in the developed countries are willing to participate in that kind of redistribution of their own wealth. "So it seems to me that it's a little unctuous to say that richer people in India should be eating less so that poorer people can eat more." Rather than attempt such a wholesale redistribution, many development efforts rest largely on encouraging agricultural growth.

Mellor cautioned that progress in feeding the people of poor countries is likely to be very slow, if it's based solely on the rate of growth achievable in the agricultural sector of those countries. In most countries, Mellor thought, one would be lucky to get a percentage point or two ahead of the population growth rate. Without wishing to suggest that hunger problems can be solved instantaneously by calling upon the resources of developed countries, Mellor nonetheless believed it true that "you can move toward a solution of low-end poverty in the world about 25 percent faster if you enlist the developed countries and their food surpluses in that process." Such a rate increase, Mellor insisted, should not be sneezed at.

Mellor therefore proposed that more attention should be given to the apparently inevitable fact that the developed countries are not going to reduce their production growth rates sufficiently to bring them into line with their rates of growth of consumption. "How can we use that gap," asked Mellor, "to solve this hunger problem faster—instead of leaving it entirely to the developing countries, which can really only generate a surplus beyond population of about 1 percent per year?"

In Mellor's opinion, there is no way to satisfy the entitlement needs of the poor in predominantly agricultural countries without developing the indigenous agriculture. In fact, the need for development of domestic agriculture—with all the linkages and multiplier effects that this has on employment and purchasing power—is the main point of his earlier remarks. Mellor's proposal, then, suggests that this growth might be increased by perhaps 25 percent, by making use of the inevitable surpluses of developed countries.

Eugene Donefer discerned a notable omission in the day's discussions, though it had been referred to several times the previous day: the relationship between militarism and development. In his introductory remarks, for instance, the colloquium codirector had noted that the amount of money spent in several weeks of global military expenditure would provide a year's worth of food, education, and shelter for the world's poor. One month before the colloquium, Donefer related, a United Nations conference had taken place on the relationship between disarmament and development. The United States, the only major country that did not attend the conference, took the position that there was, in fact, no relationship between disarmament and development. Because the subject had been repeatedly raised at the colloquium, Donefer suggested that it would be appropriate for Swaminathan to include it in his list of discussion themes. Militarism, from the level of the superpowers to that of tribal warfare, affects development, food production, and distribution around the world.

Mamdouth Gabr proposed that the communications sector be considered in conjunction with those of education and health. In Gabr's view, it should primarily be the responsibility of the cadres of scientists that exist in most countries to use the communications media to convince policymakers of the importance of food and nutrition to the overall development of the nation. Policymakers, after all, are not all unethical, but they do all seem to consider economic development to be much more important than anything else. Bilateral and international development organizations too might be able to make good use of communications media, both to convey information to the public, and to lobby decision makers.

Sunil Roy spoke up again to urge that attention not be diverted from the immediate need for emergency action to meet the needs of hunger. "This seems to me so crucial," said Roy, "that if we go away from here without having concluded something a little more concrete than has been summed up, then we will be failing in the purpose of 'Science, Ethics, and Food.'" In considering such practical measures as Mellor proposed, Roy felt that we tend to concentrate on ethical problems in the developing world, ignoring that "there's an ethical element involved right across the board in all our countries," and that this "has to be brought home to the elite elements in all our societies." Roy urged that a consensus view be formulated on what immediate measures should be taken on the subjects of additional food aid, the improved management of food aid at the direct distribution level, and food-for-work as well as work-for-food. Roy cautioned that solutions to problems of hunger will vary not only from country to country, but from area to area within a country.

S. Bruce Shearer suggested one more element that might be added to the package of points to be agreed upon concerning immediate food needs. Shearer presumed all present would agree that the Green Revolution is not a completed revolution. Despite its accomplishments, it still is concerned with a fairly macro level of farming. It therefore hasn't yet reached as many small farmers as technology would make possible if it were fully disseminated, articulated, and developed around the world, in accordance with cultural conditions in each country. Shearer noted that there had been repeated calls at the colloquium for more grass-roots involvement of people at the receiving end in making decisions about which technologies and procedures would be most useful. He pointed out that there already exist social techniques and methodologies "that go beyond the extension systems, that go beyond market research systems, that involve direct collaboration and direct participation by groups at different knowledge levels, and different power levels, and across sectors, in order to arrive at some purposeful and useful outcomes." At their most basic level, these techniques might include simple group meetings "where everybody gets a chance to put in their two cents."

Shearer opined that the "Science, Ethics, and Food" colloquium itself was indicative of the emergence of these new technologies. "I don't think that, if this symposium had been organized ten or fifteen years ago, today's session would have been held." At that time, said Shearer, it would not have been felt necessary to have a "participatory, collaborative session, and one that ends, in fact, with the word 'consensus.'"

Shlomo Reutlinger commented on a question posed by Mellor concerning a possible conflict between development and hunger or poverty relief. Noting that he was affiliated with the World Bank, Reutlinger believed he could not be accused of underestimating the need for loans and development assistance. In reality, though, he believed there was no real conflict between the two. In the first place, there exists abundant opportunity to support private employment on individual farms—particularly if some form of incentive can be devised—so that development efforts should not have to resort to public works.

Reutlinger then provided a few reflections on the shortcomings of the current international efforts to alleviate poverty and hunger. Reutlinger was dismayed to report that very little of the food-aid resources are actually used for poverty and hunger alleviation. He estimated that only 10 or 15 percent of total food aid—some $300-400 million—is used to ameliorate the problems of two hundred to four hundred million desperately poor people. If we all recognize how inadequate these funds are, he asked, why then are we not expanding these programs for planning and food development? One answer is that there is vast room for improvement in the highly inefficient and inappropriate current programs.

Much of the problem with present programs, Reutlinger felt, resides in their reliance on public distribution of food. Such systems deliver to the people perhaps only 30 percent of the value of its resources, while 70 percent of it somehow goes to feed the system of distribution. A possible way around this, Reutlinger believed, lies in the use of food vouchers or stamps. Even existing currency could be used if a voucher system proved too complicated. But in assuring that the people in need of food are able to pay for it, one must also make sure that they are provided with a medium of payment that they can allocate as they see fit. In so doing, one creates a much more efficient distribution system—perhaps preserving 80 percent to 90 percent of the aid resources. Having done that, one should then turn one's attention to the quality of the aid programs themselves. Educators should be responsible for providing good educations, and health workers should attend to health problems and nutrition education. Neither should be responsible for distributing food packages (though Reutlinger would permit them to give out food stamps). Small, private enterprises should handle the actual food.

Reutlinger detected a great need for an international financial facility to deal specifically with the problems of the poor. The IMF, after all, has a special facility to deal with debt problems and to attempt to reestablish growth where it is lacking, but there exists no facility for helping poor people. "I think such a facility should be promoted," asserted Reutlinger, "perhaps first, not with $10 billion, but with $1 billion." He estimated that half of that amount could be derived from current food aid delivered to poor countries in the form of actual food, and then sold on the open market, with the resulting funds distributed in the form of food vouchers to those needing food. Another 25 percent of the facility's costs, Reutlinger suggested, could come simply from more foreign exchange to the countries, "because when you increase the income of people, they're going to want other things as well, whether it's education facilities or health facilities; there's foreign exchange that is going to be needed." Perhaps another 25 percent could be simply contributions from the countries' own domestic resources.

Swaminathan remarked on the similarity of Reutlinger's proposal to the activities of the World Food Program, which, in accordance with Reutlinger's ideas, has attempted to adjust its imbalance of commodities and cash by instituting a rural employment program. Reutlinger responded that his proposal was being discussed with the World Food Program.

Noting the brief time remaining for discussions, Swaminathan urged that the group turn its attention to the questions of what dimensions hunger had assumed, and what were the various approaches that might be taken in reducing hunger. He solicited the views of representatives of private industry and of a fellow of the African Academy of Science.

G. E. Marler, of IBM's World Trade division, described IBM's partnership programs with universities, as well as with agricultural and other research institutes, worldwide. One development that he felt should be highlighted was the formation of computer networks that link the mainframe computers of key universities, so that professors and students equipped with terminals can exchange information, research data, and documents. These networks have now spread through twenty-five countries around the world, including Canada, the United States, and others in Western Europe, the Middle East, and the Far East. The possibilities of this kind of network in food research, agriculture, aid, and government organizations are indicated by the system currently in place at IBM offices now. Marler explained that his office is linked to 400,000 IBM employees in 120 countries, providing him with the capability to "move documents and information effortlessly from Pakistan or India or really throughout the world at a moment's notice." This kind of system greatly enhances the productivity of scientist-to-scientist interaction, Marler remarked. Marler also described a second fairly recent development in the use of the microcomputer, particularly the laptop PC, in fieldwork, in child health, in public health, agricultural research, and disaster relief. In the field of disaster relief, the personal computer has recently been tied by volunteers into the worldwide HAM radio network.

Thomas Odhiambo provided a number of comments on the subject of science and technology education, particularly as it relates to long-term solutions to Africa's food problems. Science, Odhiambo insisted, "must be part and parcel of the education and culture of the people themselves." Currently, he noted, people in Africa, particularly the new leaders, adhere to the notion that science and culture are separate entities. When political leaders speak of applying science to national development, Odhiambo lamented, "they are simply talking." Rarely are such ideas put into effect through national policies, or in actual practice. "So we have a major educational and sensitization mission to bring the question of science and technology into real life . . . to make the political leadership understand that, in the twenty-first century, one will have to lead by the brain, and that much of the brain will be scientific."

This way of thinking has deep roots, said Odhiambo. At independence, for instance, the major way in which the developed world sought to assist Africa was by encouraging the simple adaptation of Western research results, rather than by encouraging local research efforts. This attitude—that independent research is not necessary in Africa—is now deeply ingrained in the leadership of the continent. It does not appear to be as prevalent in other regions of the developing world.

In order to change this attitude, and make science a part of ordinary African life, Odhiambo believed dramatic changes will have to be made in education, beginning at the primary level. As Borlaug pointed out, said Odhiambo, primary education in Africa is "a mess." Currently, children in African primary schools are well versed in European folk music and fairy tales, but their science education is thirty or forty years behind the times. "We have to start a wholly new tradition of bringing modern science into the cultural parlance of the African child and the African mother," he insisted.

At higher education levels, Odhiambo declared, a serious problem exists as well. As Borlaug had explained earlier, one of the major reasons the Green Revolution succeeded in India was that agricultural education at the university level in that country was highly advanced. Such is not the case in Africa, noted Odhiambo. It is particularly lamentable that, even now, many years after independence, most African countries still do not have Ph.D.-level training in agriculture, and many do not even have schools of agriculture. None has a postdoctoral program in agriculture. No real leadership exists, therefore, in agricultural research and development in extension work.

Odhiambo admitted that he was painting a somewhat simplistic picture, for there are pockets of strength here and there in Africa. Nigeria, for instance, has some remarkable scientific talent, though much of it is not in the agricultural field, and, unfortunately, much of it has actually left the country. This is true also of Ghana and Uganda. The trained agricultural and agronomist leadership is only now returning to Zimbabwe. In Odhiambo's view, "It's important to know that we have a ten- or fifteen-year process of bringing our agricultural faculties, our agricultural colleges, and even agricultural universities to a level where they can make a major and continuing impact in agriculture, not only through production, but also all the way to marketing, processing, and food science and technology as a whole."

The picture is not uniformly gloomy, though, said Odhiambo. A positive step has been taken in the establishment of an African Academy of Sciences. This organization not only honors those who have made signal achievements but also endeavors to mobilize development talent—in technology as well as in the hard sciences and social sciences.

Another encouraging development Odhiambo has observed is the recent establishment of a number of private universities. For the first time, Africa is experiencing a competitive situation in higher education. This may not be unusual in other lands, but in Africa it has traditionally been taken for granted that higher education is the sole preserve of the government. With the rise of private colleges, commissions of higher education have now been established in Ghana, Nigeria, and Kenya to assure that these new institutions meet at least the standards maintained at government universities. "Our hope is that competition will ensure that the private universities are actually doing a little better than the public universities, for . . . we are now beginning to think in terms of quality." African education has reached a point now, Odhiambo believed, where it should consider whether the experts it produces are world-class. Are the scientists, technologists, institutional managers, and marketing specialists in a position where they can compete with the best anywhere in the world? "This will be the question for the next decade, but we are making a beginning already," asserted Odhiambo.

John Stoval, chief of the research division at the Board for Food and Agricultural Development, was anxious to note that we need not start from scratch in addressing food problems. On the contrary, "we do know what needs to be done. We know that the indigenous capacity has to be developed in the developing countries, and we can't do it overnight . . . but I think we know how to do it. And we have the means to do so. We have the know-how and the capacity." What Stoval felt we lack is the public support—not only in the developed countries, but in the developing countries themselves. Resources and commitment therefore need to be mobilized in both the developing countries and the developed countries, he asserted.

Jessica Mathews took the floor to address the issue of science transfer. Brady, she recalled, had earlier quite pointedly underlined the importance of setting our research and development priorities, and Mathews wished to indicate what she thought the key priority should be on any agenda for action that grew out of the colloquium discussions. There seemed to be general agreement, she observed, that there was no shortage of food on the global level. There are, however, a great many hungry people, and their numbers are growing. They are hungry, she said, because they are outside the market. The colloquium discussions had raised a number of problems associated with the redistribution of surplus foods, noted Mathews, and it is clear that this alone will not solve the problem. While the problem involves increasing production, it must be increased for the people who are hungry, where they exist, and, to a large degree, by them. In general, the hungry live on land that is fragile. A great many of them live on steep, erosion-prone slopes. Many live in countries where the reach of the central government hardly extends fifty miles beyond the capital city. A great many of them live far beyond roads, far beyond access to credit or to reliable deliveries of purchased inputs. Mathews would therefore suggest that the research and development priority for the international community and for the developing countries, as well as for the international centers, should be in developing technologies appropriate to these conditions. Initially, this is not a matter of highly intensive production, but rather is about technologies that can provide simply a balanced-nutrition, subsistence living. Eventually, it could aim for production of small additional amounts of products that could be sold on the local, regional, and, ultimately, national markets. This might at some point pose a challenge to Western agriculture. Although Western agriculture is perceived as the world's great success story, it has relied on a totally different model, based on temperate ecology, temperate soils, and industrial country economics, involving heavy reliance on purchased inputs. Under the different conditions in developing countries, reflected Mathews, "I really wonder how much Western agriculture has to offer." Certainly, one danger in this sort of agriculture is that it might induce the kind of ecological damage—such as erosion or salinization—that later either reduces or destroys agricultural productivity. Once we have recognized the necessity of developing technologies appropriate to local conditions, Mathews admitted that the problem would then remain one of finding the expertise needed to develop it.

An unidentified speaker remarked that there seemed to be no disagreement on the need for advanced scientific and technical training in developing countries. In the 1960s, the Rockefeller Foundaton recognized this and embarked on efforts to develop leadership and university education in a number of Third World countries, by financing fellowships for advanced training in designated fields. When the United Nations University was established in 1975, it was recognized that it could make a major contribution to institution building in developing countries because, unlike other United Nations organizations, its fellowship program was not limited to candidates proposed by governments. It could go directly to institutions in developing countries and find out what their needs were and who the people were that the institutions needed trained. On this basis, the United Nations University has now awarded over a thousand fellowships—over 60 percent of them related to food or nutrition—and has had a profound influence on the development of many institutions. A great deal more could be done in this area by the international community, bilateral assistance agencies, foundations, private organizations, and even by individual universities. By providing advanced training and improving university education, the speaker maintained, one can make a very important contribution to the food, nutrition, and health aspects of developing countries, as well as to their overall social and economic development. This training should be conceived in terms that are appropriate to conditions and needs in the individual developing countries. A recommendation in this area from the colloquium group, it was suggested, could make a great deal of difference.

Robert Kates provided an analysis of the discussion thus far. By attempting to examine different sectors of the problems, Kates felt that the group had inadvertently adopted a common fallacy of considering science as something we do, and policy as something we argue about. Kates had drawn up a list of issues raised during the course of the colloquium, and he felt it was significant that the issues seemed to fall into pairs of antithetical choices, among them: whether to build agriculture, or whether to build the general economy; whether to raise food, or whether to buy food; whether to give food, or whether to give money; whether to use food availability as a model, or whether to use entitlement as a model; whether to urge self-sufficiency, or whether to urge reliance on international trade and surpluses; whether this is a time of meanness, or a time of promise; and, the "most difficult" question of all—posed earlier by Robert Paarlberg and since then conveniently skirted—whether one can create a just food system in an unjust world. Kates emphasized that these differences need to be confronted and not glossed over in an attempt to arrive at consensus.

Research on a number of these questions is actually already well underway, said Kates, at such places as the International Food Policy Research Institute and the World Bank, and a growing body of information exists to inform the debate. Even on the question of whether a just food system is possible in an unjust world, "the fact is that hunger has been reduced in societies that are quite unjust, and we can look at those examples and ask how that was done, and what does that mean, and what shall we learn."

Kates returned to the questions of ethics and education, with some remarks on the type of training that should be offered to students in developed countries. Interest in issues of hunger, Kates noted, seems to be increasing among students today. In his new undergraduate course on hunger at Harvard, for instance, Amartya Sen had fifty-five students enrolled, and one hundred auditing. In Brown University's course on the subject, fifty were enrolled, and seventy sat in. Each student enrolled in the course at Brown was required to perform work in a food kitchen in order to have some direct exposure to hunger. Of the six thousand students at Brown, 1,200 are engaged in some form of public service. "It seems to me," said Kates, "that perhaps the time of meanness is coming to an end. We see reflected in our own youth the beginning of the new, still very tender, still very fragile signs of a renewal, of a time of promise, and we in higher education in science in the developed world should do everything we can to encourage that."

Kates concluded with an open invitation to attend a hunger briefing and research exchange at Brown University the following April—"the first of a series of conversations between the research community and program managers, private voluntary organizations, and hunger activists." Kates also invited colloquium participants to join with forty groups around the world in a Hunger Research and Food Systems Exchange, through which findings are shared around the world, by mail.

Nyle Brady was abashed to find himself the last person called upon to speak, but did venture some comments on the subjects of training and "misconceptions" about technology. The misconceptions among African leaders about the possibilities of science and technology, mentioned by Odhiambo, Brady asserted, were quite similar to those he had noticed in American leaders. They seem to believe that all the necessary technology already has been developed, and "all we've got to do is take the things that we've learned in the United States, and take them over to those poor, ignorant Africans and let them put it to work, and then they'll take care of it." We have got to make it clear to our leaders that the technology does not exist, insisted Brady, and it also has to be made clear that, in Africa, the technology will have to be developed there, "and to the degree it possibly can be, by Africans."

On the subject of university training programs, Brady pointed out that there is more involved than simply bringing students from developing countries to the United States to study. Brady took issue with Odhiambo about the absence of trained educators in Africa. In many countries, Brady insisted, Africa's best-trained scientists are at the universities, but most of them are not able to devote their attention to the crucial matters addressed by Odhiambo, because their institutions have not yet been developed and supported adequately to make use of the training of these scientists in areas where it is badly needed. Many of these scientists, Brady observed, are "seeking money to continue the work that they did in the United States or Europe on their theses, that may have no relationship to Africans' problems." Greater attention therefore needs to be given to training people in areas that are relevant to the problems of their region. As in Asian countries such as India, the Philippines, Indonesia, and Thailand, priorities must be set. Brady did not believe that Africa can afford to develop forty institutions of international stature, but it might be possible to develop five or six first-class agricultural institutions. Asked recently by the vice chancellor of the Kenya University College for advice on how to develop his institution, Brady suggested he first formulate a twenty-year plan, outlining the kind of institution desired and emphasizing the research component. With this plan in hand, support could be sought from the national government, the World Bank, the United States, Germany, or institutions abroad willing to undertake long-term cooperation.

Swaminathan announced that time for discussions at the colloquium had come to an end. At the beginning of the session, he recalled, he had outlined three major issues to be discussed: how to make the concept of "the right to food" a reality; how to develop national capacities for

Wilton Dillon and M. S. Swaminathan

science and technology; and how to develop sustainability. Though there had not, unfortunately, been time to address the third of these issues, Swaminathan referred participants to a recent World Commission report that had dealt with it in some detail. Many of the basic issues concerning science and technology—notably the role of private industry in adapting relevant new technologies—had been at least mentioned at the colloquium, Swaminathan noted. Because there were a large number of issues that could not be discussed, he proposed that the Smithsonian and General Foods might assemble a small working group of participants to examine the results of the colloquium, "to sift it and develop some kind of a follow-up action plan."

Wilton S. Dillon, the director of the Smithsonian's Office of Interdisciplinary Studies, had the last word, extending a vote of thanks to those "individuals and groups who have made this extraordinary month of activities in the Smithsonian come to a very special climax with the events of yesterday and today." As the first major gathering in the Smithsonian's new International Center, Dillon suggested that it could be considered historic, as a model for bringing together "individuals from the knowledge industry, and those people who care about people, and those who wish to ameliorate the human condition." Dillon expressed particular thanks to the representatives of the General Foods Corporation, who, together with Winrock International, worked with the Smithsonian Institution on planning and supporting the colloquium.

Dillon remarked that, as an anthropologist, he had long been trying to apply the human sciences to an understanding of various kinds of reciprocity chains. War and peace, for example, are parts of such chains, the links of which are determined by whether one exchanges revenge for revenge, love for love, or simply cooperation for cooperation. Like cooperation and love, food is a commodity that we seek and want to share, out of a sense of what Dillon termed "enlightened self-interest." Referring to the three human universals identified by the anthropologist Marcel Mauss—the obligation to give, to receive, and to repay—Dillon suggested that the repayment process be further investigated at future such gatherings. What do those who control food as a major commodity receive back from those who are recipients of it? How early in human life can food nutrients create the qualities we look for in the mind? Most of what we call civilization, Dillon observed, will ultimately depend upon this bottom-line supply of food. In the politics of food, he suggested, it appears that René Dubose's advice to "think globally and act locally" will lead to the most practical efforts aimed at maintaining both this supply and some semblance of civilization.

APPENDIX A

Our Common Agricultural Future

Acceptance Address on the Occasion of the Presentation of the General Foods World Food Prize, 6 October 1987

I am grateful, both to those who conceived this prize and to those who decided to award it to me. I am aware that many in our spaceship Earth are more worthy of this recognition. I am equally aware that, although I was singled out for the recognition, I could have accomplished little in my life without the generous support of many other men and women.

My late mother and an uncle gave a sense of mission and direction to my academic and professional life after I lost my father when I was eleven. My wife, Mina, has been my principal guide and source of inspiration. She has encouraged my work, at considerable sacrifice to her own professional and personal life. My daughters Soumya, Madhura, and Nitya have helped me remain young in my thinking and bridge the generation gap in terms of values and aspirations.

My scientific colleagues have been generous in sharing their ideas, material, and suggestions—colleagues at the Central Rice Research Institute, Cuttack; the Indian Agricultural Research Institute, New Delhi; the Indian Council of Agricultural Research; the International Rice Research Institute; and in numerous universities and institutions in India and abroad. Over the years of my professional life, I have observed that progress in plant breeding was most positive when we shifted attention from individual plant characteristics to population performance. The semidwarf wheat varieties that Dr. Norman E. Borlaug sent to India in 1963 were not impressive if examined plant by plant, but were superb when examined collectively. Those varieties triggered India's wheat revolution. In other words, the total yield per hectare was a result of collective performance rather than of individual excellence.

M. S. Swaminathan

Similarly, the contributions for which I am receiving credit are the products of teamwork and national and international collaboration. Two individual scientists to whom I am most indebted for inspiration and guidance are Dr. B. P. Pal in India, and Dr. Norman Borlaug in Mexico.

Development of improved technology alone cannot help farmers grow more food unless the diffusion of that technology is facilitated by public policy. Here again, I have fortunately had access to political visionaries: Mr. C. Subramaniam, who laid the foundation for scientific transformation of Indian agriculture at a time when the country was described by foreign observers as leading "a ship-to-mouth existence"; and the late Prime Minister of India, Indira Gandhi, whose faith in both science and food reserves is responsible for India now being able to manage severe drought and floods without importing food.

Farmers have given me guidance and encouragement that have proved extremely important in my work. I was fortunate to learn from Indian farmers rather early in life that they will respond even more to action-based communication than to lectures. For example, colleagues and I organized a seed village in Delhi State in 1964 to accelerate the multiplication of pure seeds of some of the Mexican semidwarf wheats. Farmers were rather reluctant to respond initially, but they soon became enthusiastic. When I asked a farmer why their attitude had changed, he said, "we noticed that you always visit the project on Sundays. That means you are not being paid for this job—so we know that you are here out of a conviction that what you recommend is for our welfare." Such farm education helped me to design an effective national demonstration program in the fields of small farmers, based on the principle that "seeing is believing."

I have had the good fortune of working with scientists from many nations. Numerous teachers, scholars, students, political leaders, administrators, extension workers, journalists, development communicators, and farm laborers have helped and influenced me. My brothers and sister, parents-in-law, sons-in-law, nephews, and other relatives, friends, and an invisible college of correspondents around the world have been pillars of strength. I am particularly grateful to critics, because they help me to analyze my shortcomings and to improve. The institutional support I have enjoyed during the last forty years has been a dominant factor in my scientific, organizational, and administrative work.

To all of you, I take this opportunity to say, sincerely, "Thank you."

Because I have derived immense pleasure and power from the partnerships, I have—with the support of my wife and family—decided to use the cash award associated with this honor to promote a small initiative in fostering meaningful farmer-scientist partnership in the development of an integrated approach to biological and social engineering, as applied to technology development and diffusion under small farm conditions.

May I now take a little of your time to explain my personal philosophy of agriculture.

Once, at his breakfast, Martin Luther King said:

> All life is interrelated. . . . We are caught in an inescapable network of mutuality. . . . Before you finish eating breakfast in the morning, you've depended on more than half the world. This is the way our universe is structured; this is its interrelated quality. We aren't going to have peace on earth until we recognize this basic fact of the interrelated structure of all reality.

In no other area of human need and endeavor is there so much global interdependence as in agriculture. Yet the urban public seldom recognizes that we live in this world as guests of green plants and of the farmers who cultivate them.

Experience shows that countries that take farmers and farming for granted come to grief sooner or later. This year's weather has been erratic—severe drought threatens some areas, floods have washed away crops in others. Its adverse impact on food production in densely populated South and Southeast Asian countries is a timely warning. Complacency is creeping into the thinking of many planners and political leaders as they establish the priority accorded to the farm sector in national development plans. We have no time to relax on the food production front, as Dr. Borlaug often reminds us.

Globally speaking, reserves of food grains, milk powder, and butter are growing daily. Simultaneously, the number of children, women, and men who go to bed hungry is also increasing. Why have all our intellectual, technological, financial, and spiritual resources failed to find a solution to this age-old irony? Why?

Socrates probably gave us the answer: "Nobody is qualified to become a statesman who is ignorant of the problems of wheat." If statesmen who determine national policies and priorities would all become conversant with food production and equitable distribution, hunger could be made a problem of the past sooner than otherwise would be possible. That was the hope expressed by President Franklin D. Roosevelt when he convened the 1945 conference that led to the birth of the Food and Agriculture Organization of the United Nations. Since 1945, many global congresses have been held and as many resolutions have been made calling for the elimination of hunger and its mother, poverty. But unfortunately, the situation worsens.

Sixty years ago, Mahatma Gandhi said:

> It would conduce to global progress and save a great deal of time and trouble if we cultivate the habit of never supporting resolutions—either by speaking or voting for them—if we had not either the intention or the ability to carry them out.

I wish that this would become the guiding principle for government representatives speaking in national and international forums, so that we do not have to listen to delegates eternally talking poor but living rich.

China, which has more than 20 percent of the world's population, has ensured basic food security for all its people. Here is an example of how political will, coupled with appropriate development strategies, can give every child an opportunity to realize his or her potential. India, with the next largest population, has shown how agricultural production can be increased to desirable levels; but millions of people still suffer from undernutrition due to inadequate purchasing power. India has, however, avoided famines, thanks to the policy of maintaining substantial food grain reserves coupled with operating a large public distribution system and rural works program.

Meanwhile, population and pressure on the supporting capacity of the land continue to grow—4.8 billion of the expected global population of 6.1 billion will be living in developing countries in the year 2000. The famine of jobs, which is already serious in many predominantly agricultural countries, is likely to increase in quantitative and qualitative dimensions. As a consequence, inadequate purchasing power will grow in importance as the root cause of undernutrition.

The World Commission on Environment and Development, in its report "Our Common Future," released by Norwegian Prime Minister

Gro Harlem Brundtland in April this year, has shown how ecological, economic, and ethical imperatives link the lives of all inhabitants of Mother Earth. Still, we witness increased polarization and fragmentation in terms of country, religion, language, and color.

Genetically, diversity is our greatest endowment. Within the enormous diversity we see in biological populations, however, genetics shows an underlying unity, in terms of the chemical substance and mechanism of heredity. How can we reconcile these differences in human behavior and identify common action points that could bring us together?

The elimination of hunger and its real cause, poverty, should be at the top of the human agenda for common action. Unfortunately, the well fed do not seem to be very concerned with the hunger of other people, although humanitarian responses to tragedies have time and again been manifested, as in the Ethiopian famine. Most people fear that "if others get more, I will get less." We need to accept the idea that the survival of the human species depends on a harmonious relationship with each other and with the earth on which we live.

How can we bring about this change in attitude and behavior? If we are really serious about eliminating hunger, we have to overcome the resistances to a change in systems that can bring this about. Economic interests, the sociopolitical interests of small groups, and sometimes sheer ignorance or indifference combine to form a political will that makes the poor remain poor.

The roots of such attitudes are many and varied, but let me say that the taproot is fear—a fear of having to share power and resources. We therefore need to show that helping the weak to become strong solidifies the whole community.

Another factor is the failure to identify ourselves as members of a global family. A Westerner may be talking about the "starving Ethiopians"; an Indian, like myself, may be talking about "the poor," "the tribals," or "the illiterate women." Each term denotes a specific group and is useful for certain purposes. But it tends to stop us from thinking about them as members of the same human family. The concern is simply not there.

How then can we generate concern for other people with economic and social handicaps? How can we help transform this concern into meaningful action? Concentrating on nutrition security for all people at all times, let me refer to three areas where people everywhere can and should work together.

First, I would like to see high priority accorded to fighting what I may call the "ecological fire" raging in most parts of the world. Developing countries are ravaged by this fire, through such mechanisms as deforestation, soil erosion, desertification, water pollution, and overpopulation. Developed countries suffer equally, through atmospheric pollution, acid rain, contamination of water through toxic wastes, and environmental mutagens and carcinogens.

Certain phenomena, such as loss of biological diversity, destruction of the ozone layer, nuclear fallout, and potential climatic changes as a result of carbon dioxide accumulation in the atmosphere and ocean-warming will affect us all, regardless of the problem's geographic origin. Today, we are concerned with fighting fires in buildings and forests. Ironically, we are not fully aware of the vast dimension and potential impact of the ecological fires that affect our basic life-support systems of land, water, flora, fauna, and atmosphere. Some hopeful signs are fortunately emerging. The protocol for limiting the use of ozone-destroying chemicals signed recently in Montreal by twenty-four nations, and the development of a time-bound tropical forestry action plan, are good examples of this hopeful trend. We need to strengthen this trend. Responses to the environmental challenges we face should include measures for maintaining biological diversity and ecological processes. Planning and managing irrigation systems, promoting natural forest growth, and conserving the entire system of production from upstream forest to downstream fishery are vital for an enduring food and nutrition security. The ecological fires that can destroy the livelihood security of generations yet to be born cannot be put out in a day, or a month, or even a year. Extinguishing them needs long-term commitment and innovative approaches. Ecological relief to undertake afforestation, soil conservation, and other measures often is not easily available in seasons when there is adequate moisture in the soil. Fifteen years ago I pleaded for a good weather code for undertaking ecological rehabilitation programs in years with normal rainfall. Unfortunately, national and international aid disappears when the rains come.

Equally important is the promotion of what I would call a "symphonic agricultural system," based on integrated principles of ecological sustainability, economic viability, and equity. The various links in the production-marketing-consumption chain of a symphonic agricultural research and development program are designed to promote growth without loss of stability. Quite frequently, the concept of a sustainable agricultural production system is used to plead for the preservation of the status quo or to revert to old production technologies. A dynamic concept of sustainability is necessary to help us meet the needs of an expanding population while maintaining and enriching the natural resource base.

In most developing countries, a vertical growth in productivity and a higher intensity of cropping are the two major pathways through which the additional food needed will have to be produced. Many of them, fortunately, have the capacity to increase both yields and cropping intensity with current technology levels. For example, given an optimum blend of three elements—viable technology, efficient services, and supportive government policies—we can double the average yields in several countries with chronic food shortages with the technologies now available. Dr. Borlaug and his colleagues are demonstrating this in several food-deficit African nations.

The president of the South Commission and former president of Tanzania, Dr. Julius Nyerere, once remarked: "Many expatriate experts come to me and say that Tanzania has a rich productive potential. I tell them I am tired of hearing about the production potential of my country. What I want is production." While external help can be useful, and is often necessary, converting untapped production potential into actual production is largely in the hands of the government and people of every country. Unfortunately, governments in developing countries tend to exhibit a strong urban bias in their resource allocation process. Rural areas get little attention as far as basic provisions and infrastructure are concerned. Mahatma Gandhi once said that the most serious form of brain drain that occurs in India is the migration of skilled and educated persons from the village to the city. Fortunately, new technologies that require both brawn and brain can provide intellectual and economic satisfaction to young farmers. They, therefore, provide an opportunity to retain educated youth in villages. Youth and the poor constitute the two genuine majorities in the Third World. Unless youth can feel excited about the opportunities now available for a technological transformation of rural professions, the fate of the poor will be bleak.

The mass media term "miracle seeds" created a misleading impression that agricultural progress can be achieved through miracles. There are not short cuts to progress in agriculture or aquaculture. Above all, people should not be misled into thinking either that modern inputs are not needed for output or that biotechnology will solve all our problems. Output cannot be increased without the appropriate inputs. Tropical soils, often hungry and thirsty, support excellent crops if they are provided with water and nutrients. Therefore, we should aim to reduce not the essential inputs to the crops, but rather the cost involved. Recent research in the International Rice Research Institute and other institutes has shown that cost reduction without yield reduction can be achieved by using purchased inputs efficiently and by substituting farm-grown inputs for some of the market-purchased ones. This is an area where there is much scope for an intelligent integration of traditional and emerging technologies.

Prices in international markets for primary agricultural commodities and opportunities for trade greatly influence the agricultural health of many developing countries. Some hopeful trends are now emerging that could lead to stability and equity in international trade in agricultural products. If developing countries are not given a fair deal in agricultural trade, many of them will never be able to remove the severe debt burden that is crippling them economically and politically.

John Mellor, in his lecture at the colloquium held today, stressed that the best way to improve the status of the world's poor is through the transfer of scientific resources. This is because a dynamic agricultural production program can be initiated and sustained only with the help of a dynamic national research system. The International Agricultural Research Centers (IARCs), supported by the Consultative Group on International Agricultural Research (CGIAR), have helped many nations buy time by providing the needed backstopping, in both research and training. The stronger a national research system, the greater is the benefit derived from an IARC. Unfortunately, many developing countries do not realize the pivotal role of science and technology in converting the natural endowment of a country into wealth meaningful to its people. Research gets low priority in resource allocation, and research institutions are ill equipped and inadequately funded. Consequently, good scientists go to the better endowed institutions in developed countries. Apart from better working conditions, the growing gap in pay, privileges, and perquisites between scientists employed in UN organizations, IARCs, and advanced institutions in developed countries and scientists in national research systems of developing nations is becoming a major cause of brain drain. A low-cost mechanism on the Peace Corps model will have to be developed to fill critical gaps in technological expertise, where needed.

The trend toward total privatization of plant breeding research in developed countries and the ever-widening scope of patent rights also cause concern in developing countries. From such a concern arises the recent move in FAO to promote, in developing countries rich in crop-genetic resources, the concept of farmers' rights to compensate for the breeders' rights prevalent in developed nations. Seeds, which from the dawn of agriculture have known no political frontiers until now, may soon be required to have, in addition to phytosanitary certificates, "passports" to be issued only upon substantial payment. Thus technology, as well as trade, is becoming a source of dispute and discord. It is sad that agriculture, which could be a major instrument for international cooperation, is becoming another source of polarization. It is in this context that the strengthening of a nonpolitical and noncommercial scientific organization like the CGIAR assumes urgency.

A third area that requires attention is an analysis of the reasons for our inability to achieve freedom from hunger. The problem needs understanding both at the production and consumption levels.

Taking production first, political leaders often tend to take a short-term view of essentially long-range problems. They are therefore interested in "crash agricultural production programs" that are launched with a good deal of political fanfare. Such programs generally ignore the human beings whose toil is vital to their success. It is not surprising that many of them collapse—after having wasted scarce resources.

I have come across many agricultural plans developed by agricultural departments in Asia and Africa where the word "farmer" is not mentioned, even by mistake. And these are documents related to raising production! All the other ingredients are there—credit, seeds, irrigation, fertilizers, pesticides, and farm implements—but not the human being who will use all these inputs. And when the term "farmer" is finally used, it is mostly taken to refer to men, and seldom to women farmers and farm labor. Yet we suspect women played a leading part in the domestication of plants more than twelve thousand years ago. In most developing countries, they continue to perform key functions in seed selection, food production, postharvest conservation, and organic recycling. Also, the poorer the household the greater the need for women to have access to independent income.

Yield targets are frequently set in most government plans, but seldom do we see a minimum income target for farming families. Public and private sector employees are always concerned with their net take-home pay, continuously adjusted against inflation. But the same officials, when formulating policies for the farm sector, do not think that small farmers in the self-employed sector deserve a minimum take-home pay. A minimum income target for families engaged in rural occupations also ensures adequate resource flow to the villages and helps correct imbalance in the terms of trade between the farm and nonfarm sectors. We need a new deal for the rural self-employed in the Third World whose livelihood security depends mainly on land- and water-based occupations.

In this context, I would like to pay tribute to President Corazon C. Aquino of the Philippines, whose government has decided that the aim of agricultural progress is to increase in real terms the income of Filipino farmers—from about $60 a month at present to $100 per month by 1992 (in U.S. dollars). Unless we can reorient the thinking of governments along these lines, we will always have politicians and professionals talking about the untapped potential of their countries but not doing anything to realize that potential.

The livelihood security of the poor in rural areas depends upon diversified opportunities for employment both on the farm and in off-farm sectors. To undertake this task on a scale that matches needs, more investment is essential in rural areas. FAO has estimated that an investment of $1500 billion (in U.S. dollars, 1980 prices) will be needed up to the year 2000, if developing countries are to achieve the necessary growth rates in agriculture. The Brundtland Commission has dealt with the interrelationships among ecological health, food security, and arms expenditure. The global expense for arms is now estimated at $2 million per minute, or nearly $20 billion a week. In spite of such an expenditure, international surpluses in terms of cash, commodities, and commercial technology are abundant in developed countries. A recent study by a group of economists at the World Institute for Development Economics Research (WIDER) points out that during the next five years, Japan alone could easily transfer resources worth $125 billion (in U.S. dollars) to developing nations. Resources can thus become available if there is a political will to help the poor. Father Byron rightly stressed that political will by itself will have little impact unless there is also a political way to accomplish the desired goal.

We should constantly remind ourselves that hunger robs millions of individuals of the opportunity to lead fuller lives. And today, although we know that there is no valid excuse for hunger to exist, we still do not know how to remove this stigma from our civilization.

May I suggest three major lines of action to supplement present efforts.

First, we need to make people aware of the fact that we can create a truly joyful world, where not a single person spends the night hungry. How can we do this on a scale that can have an impact? Television is obviously the communication medium of choice. If about two billion people can watch the Olympic Games or the World Cup football tournament on TV around the world, can we not let them see a "Freedom from Hunger Olympics," where countries and organizations demonstrate the methods they have used to provide physical and economic access to food to everyone? May I appeal to the General Foods Fund to take the lead in organizing this World Without Hunger Olympics? A global event like this might be organized every two years, with the help of national and international organizations committed to promoting ecological and nutritional security. Among them are the Better World Society, the Hunger Project, the Worldwide Fund for Nature, the International Union for the Conservation of Nature and Natural Resources, the International Council of Scientific Unions, the UN/FAO World Food Program, UNICEF, and many others.

A second suggestion is the development of a mechanism through which scientists and farmers from different countries can share experience and know-how. Currently, there are "peasant to peasant" organizations. Similarly, scientists have many opportunities to meet and exchange ideas. The communication revolution we are witnessing now will increasingly facilitate continuous intellectual interaction across continents and reduce the need for formal conferences. What is missing is a mechanism for promoting joint scientist-farmer undertakings. If agricultural research becomes a joint-sector activity involving scientists and farmers, the growing gap between knowing and doing can be bridged. Also, the pace of development and diffusion of location-specific technologies can be accelerated.

Finally, we need intervention programs tailored to meet the specific needs of countries where hunger is prevalent. Professor Amartya Sen has shown in several of his books and articles how well-designed public intervention measures that enhance entitlements can reduce or eliminate hunger and expand life expectancy. Agrarian reform and employment guarantee schemes are important in this respect.

The challenge lies in developing strategies at the national level for utilizing food and commodity aid in a way that will help end the need for such aid, not perpetuate dependence and erode self-reliance. Considerable experience in this field already exists within international organizations such as the World Food Program and UNICEF as well as within many bilateral and national organizations. Based on an analysis of this experience, an action code for initiating appropriate "Food for Self-Reliance" and employment guarantee programs can be prepared.

To sum up, I plead for coordinated nongovernmental initiatives at first, generating awareness by taking advantage of the powerful mass media tools now available; second, analysis of field problems and developing solutions for them jointly by farmers and scientists; and third, action in utilizing the growing global grain surpluses for enabling all those in need of help to earn their daily bread.

We live in an age of unparalleled opportunity for promoting sustainable nutrition security. The prospect for a world without hunger is a glorious legacy given to our contemporary world by scientists and technologists; communicators and social scientists; administrators and industrialists; and workers in the factories, fields, forests, pastures, rivers, and oceans. As we depart for dinner this evening, what could be a more satisfying and joyful feeling than knowing that every other member of the human family will also go to bed after a nourishing meal? Until such a wholly attainable world becomes a reality, our task remains unfinished.

APPENDIX B

Notes on Contributors

William J. Byron, S.J.
President
The Catholic University of America
Washington, D.C. 20064

Father Byron was educated at St. Louis University (A.B. 1955, M.A. 1959) and the University of Maryland (Ph.D. 1959), where he specialized in philosophy and economics. He holds two theology degrees from Woodstock College. In 1961, he was ordained to the Roman Catholic priesthood as a member of the Jesuit Order. His previous positions include the presidency of the University of Scranton (1975-82), deanship of the College of Liberal Arts and Sciences, and associate professorship of economics, Loyola University (1973-75). Among his publications are *Toward Stewardship: An Interim Ethic of Poverty, Pollution and Power* (1975); *The Causes of World Hunger* (editor, 1982); and approximately one hundred journal, newspaper, and magazine articles on economics, social ethics, and educational issues. In addition to serving as director, advisor, and trustee of numerous education, church, health, civic, and professional organizations, he is a member of the American Economics Association, the American Society of Christian Ethics, the American Association of University Professors, the American Council on Education, and the Association of American Universities.

John W. Mellor
Director
International Food Policy Research Institute
1776 Massachusetts Avenue NW
Washington, D.C. 20036

John Mellor was educated at Cornell University (B.Sc. 1950, M.Sc. 1951, Ph.D. 1954) and Oxford University. Previously, he was chief economist for the U.S. Agency for International Development, and before that, professor of economics, agricultural economics, and Asian studies at Cornell University. Among his numerous publications are *The Economics of Agricultural Development* and *The New Economics of Growth: A Strategy for India and the Developing World*, the latter a detailed statement of his concept of an agriculture- and employment-led strategy of growth. He has also edited and contributed chapters to *Agricultural Change and Rural Poverty: Variations on a Theme by Dharm Narain* (with G. M. Desai), *Accelerating Food Production Growth in Sub-Saharan Africa* (with C. Delgado and M. Blackie), and *Agricultural Price Policy for Developing Countries* (with R. Ahmed). Dr. Mellor has served as contributing editor for *Environment*, as member of the Board of Directors of the Overseas Development Council, as fellow of the American Academy of Arts and Sciences, and of the American Agricultural Economics Association. In 1985, Dr. Mellor became the first social scientist to receive the Wihuri Foundation International Prize. For his publications and research, he has received the American Agricultural Association Award in 1967, 1978, and 1986.

Thomas R. Odhiambo
Director
International Centre of Insect Physiology and Ecology
P.O. Box 30772
Nairobi, Kenya

Thomas R. Odhiambo was educated at Makerere College in Kampala, and at Cambridge University (B.A., M.A., and Ph.D.). His research has concentrated on natural history and on insect endocrinology. He is known particularly for his work on insect reproductive biology, on which he has produced more than one hundred technical papers, as well as for his vocal concern with science policy and development in Africa. He has taught at the University of Nairobi since 1965, and has been visiting professor at many universities in Africa and India. Dr. Odhiambo is a fellow of the Kenya National Academy of Sciences, the African Academy of Sciences, the Third World Academy of Sciences, the Indian

National Academy of Science, the Italian National Academy of the 40s, the Pontifical Academy of Sciences, and the Norwegian Academy of Science and Letters. In 1979, he was awarded the Albert Einstein Medal, and in 1987 he received the African Hunger Prize.

Amartya K. Sen
Lamont University Professor
Department of Economics, Littauer Center
Department of Philosophy, Emerson Hall
Harvard University
Cambridge, Massachusetts 02138

Amartya K. Sen was educated at Presidency College in Calcutta (B.A. 1953) and Trinity College, Cambridge University (B.A. 1955, M.A. 1959, Ph.D. 1959). His research interests include famines, food economics, endemic hunger, gender bias, welfare economics, economic development, social choice theory, ethics, and social and political philosophy. Dr. Sen has previously taught at Trinity College, Cambridge; Jadavpur University in Calcutta; the London School of Economics; Cornell University (Andrew D. White Professor); and Oxford University (Drummond Professor). His principal publications include *Choice of Techniques*; *Collective Choice and Social Welfare*; *Poverty and Famines*; *Choice, Welfare and Measurement*; *Values and Developement*; and *Commodities and Capabilities*. He is a fellow of the British Academy, a fellow and past president of the Econometric Society, foreign honorary member of the American Economic Association, and president of the International Economic Association. Dr. Sen was awarded the Mahalanobis Prize in 1976, and in 1986 he was presented the Seidman Distinguished Award in Political Economy.

Monkombu Sambasivan Swaminathan
B-4/142 Safdarjang Enclave
New Delhi 110029
India

M. S. Swaminathan was educated at Coimbatore Agricultural College and Cambridge University (Ph.D. in genetics, 1952). He then worked as a research associate at the University of Wisconsin at Madison, where his research on the potato plant led to the development of a variety known as "Alaska Frostless." Joining the Indian Agricultural Research Institute in 1954, he worked until 1972 there as a wheat and rice geneticist. During the next seven years, he served as director general of the Indian Council of Agricultural Research. He was secretary of agriculture from 1979-80 and served for the next two years as a member of the planning commission of the Indian government. From 1982 to 1987, Dr. Swaminathan served as director general of the International Rice Research Institute in Los Baños, the Philippines. He also served as independent chairman of the FAO Council from 1981 to 1985. Dr. Swaminathan is widely known as the architect of India's so-called "Green Revolution" because of his leading role in the National Demonstration Programme, begun in 1964, under which high-yield varieties of wheat and rice seedlings were planted in fields of India's poorest farmers. He is a member of the Royal Society of London, the U.S. National Academy of Sciences, the USSR All-Union Academy of Agricultural Sciences, and the Royal Swedish Academy of Agriculture and Forestry, and serves on the Board of Directors of the Better World Society. In 1986, Swaminathan received the Albert Einstein World Science Award, and in October 1987, he was named the first winner of the General Foods World Food Prize.

APPENDIX C

List of Colloquium Participants

Philip Abelson
Science Advisor
American Association for the
 Advancement of Science
1333 H Street NW
Washington, D.C. 20005

Duane C. Acker
Agency Director
Directorate for Food and Agriculture
Agency for International
 Development
Room 311, SA-18
Washington, D.C. 20523

Peter Acly
Associate Manager
Sector Communications
General Foods Corporation
250 North Street
White Plains, NY 10625

C. Eugene Allen
College of Agriculture
University of Minnesota
277 Coffey Hall
1420 Eckles Avenue
St. Paul, MN 55108-1030

Kiyoshi Ashida
C. I. Mansion Hongo 2 1001
2-167, Hongo, Meito-ku
Nagoya City, 465
Japan

Nancy Bell
American Institute of Biological
 Sciences
730 11th Street NW
Washington, D.C. 20001

Filmore E. Bender
University of Maryland
1320 Symons Hall
College Park, MD 20742

Alan Berg
Nutrition Advisor
The World Bank
1818 H Street NW
Washington, D.C. 20433

Robert Bertram
Office of Agriculture
Bureau for Science and Technology
Agency for International
 Development
Washington, D.C. 20523

L. K. Bharadwaj
University of Wisconsin at
  Milwaukee
3939 North Harcourt Place
Shorewood, WI 53211

Darshan S. Bhatia
Research and Development
The Coca Cola Company
P.O. Drawer 1734
Atlanta, GA 30301

Carmen J. Blondin
Special Associate for Trade
National Marine Fisheries Service
1825 Connecticut Avenue NW
Washington, D.C. 20235

Carla Borden
Associate Director
Office of Interdisciplinary Studies
Smithsonian Institution
Washington, D.C. 20560

Peter Bourne
Global Water
1111 19th Street, Suite 400
Arlington, VA 22209

Nyle C. Brady
Senior Assistant Administrator
Bureau for Science and Technology
Agency for International
  Development
Room 4942
Washington, D.C. 20523

Ricardo Bressani
Head of Division of Food and
  Agricultural Sciences
INCAP
Carretera Roosevelt
Zona 11
Guatemala City
Guatemala

Alan Brewster
Director
World Resources Institute
1750 New York Avenue NW
Washington, D.C. 20006

David M. Brush
General Foods Fund
250 North Street, RA-7
White Plains, NY 10625

David L. Call
Dean
New York State College of
  Agriculture and Life Sciences
Cornell University
122 Roberts Hall
Ithaca, NY 14853-5901

Doris H. Calloway
Provost
Professional Schools and Colleges
University of California at Berkeley
200 California Hall
Berkeley, CA 94720

Gelia Castillo

Leopolda Castillo

David Challinor
Science Advisor to the Secretary
National Zoological Park
Education Building
Smithsonian Institution
Washington, D.C. 20560

Chia Ting Chen
Senior Industrial Hygienist
U.S. Department of Labor
Room 3718
200 Constitution Avenue NW
Washington, D.C. 20210

Phillip E. Church
Agency for International
  Development
ST/AGR
Room 403, SA-18
Washington, D.C. 20523

Jeffrey Clark
Select Committee on Hunger
U.S. House of Representatives
The Capitol
Washington, D.C. 20515

A. S. Clausi
Director, General Foods World Food Prize
General Foods Fund, Inc.
250 North Street
White Plains, NY 10625

Liu Congmeng
Agriculture Secretary
Embassy of China
2300 Connecticut Avenue NW
Washington, D.C. 20008

Ralph W. Cummings
812 Rosemont Avenue
Raleigh, NC 27607

Dana G. Dalrymple
Office of Agriculture
Bureau for Science and Technology
Agency for International Development
Washington, D.C. 20523

Karen Darling
Assistant Secretary
U.S. Department of Agriculture
Room 228, West Administration Building
14th Street S.W. and Independence Avenue
Washington, D.C. 20250

J. E. Dutra DeOliveira
Faculty of Medicine
Sao Paolo, Brazil

John Dixon
Vice President for Technology
The Pillsbury Company
311 Second Street SE
Minneapolis, MN 55414

Eugene Donefer
McGill University
Montreal, Quebec
Canada

Sisir K. Dutta
8841 Tuckerman Lane
Rockville, MD 20854

Donald N. Duvick
Crop Sciences Society of America
Pioneer Hi-Bred International, Inc.
700 Capital Square
400 Locust
Des Moines, IA 50309

Nicolaos Efstathiadis
Counselor for Agricultural Affairs
Embassy of Greece
1636 Connecticut Avenue NW
Washington, D.C. 20009

Gretchen Gayle Ellsworth
Assistant Director for Public Programs
National Zoological Park
Education Building
Smithsonian Institution
Washington, D.C. 20560

Daniel Fairbanks
4312 East Glenn Street
Tucson, AZ 85712

Curtis Farrar
Executive Secretary
Consultative Group for International Agricultural Research
1818 H Street NW
Washington, D.C. 20433

Ashley Files
Food and Nutrition Service Specialist
New Executive Office Building, Room 7026
726 Jackson Place NW
Washington, D.C. 20503

Debbie Fish
National 4-H Council
7100 Connecticut Avenue
Chevy Chase, MD 20815

E. M. Foster
Director
Food Research Institute
University of Wisconsin
1925 Willow Drive
Madison, WI 53706

Mamdouth Gabr
President
International Union of Nutrition
 Sciences
162 Tahreer Street
Cairo
Egypt

Vera Gathright
Liaison Officer
International Fund for Agricultural
 Development
1889 F Street NW
Washington, D.C. 20006

Kenneth A. Gilles
Assistant Secretary for Marketing and
 Inspection Service
U.S. Department of Agriculture
Room 228-W
14th Street SW at Independence
 Avenue
Washington, D.C. 20250

Jamie Godberg

C. Gopalan
Director General
The Nutrition Foundation of India
B-37, Gulmohar Park
New Delhi - 110049
India

E. J. Guardia
General Foods Corporation
250 North Street, T12-1
White Plains, NY 10625

Mary Ann Haley
Peace Corps
806 Connecticut Avenue NW, Suite
 1107
Washington, D.C. 20526

Richard M. Harley
Harvard Institute for International
 Development
One Eliot Street
Cambridge, MA 02138

Robert D. Havener
President
Winrock International
Route 3, Petit Jean Mountain
Morrilton, AR 72110

Hector Herrera
1889 F Street NW
Washington, D.C. 20006

Alden Hickman
Executive Director
Heifer Project International
P.O. Box 808
Little Rock, AR 72203

Brenda C. Higgins
Acting Executive Director
League for International Food
 Education
915 15th Street NW
Washington, D.C. 20005

Rachelle Hollander
National Science Foundation
Room 312
Washington, D.C. 20550

Paul Hopper
Vice President
General Foods Corporation
250 North Street
White Plains, NY 10625

Barbara Huddleston
Chief, Food Security and Information
 Service
Food and Agriculture Organization
Via delle Terme di Caracalla
1-00100 Rome
Italy

Joseph H. Hulse
Vice President, Research Program
International Development Research
 Center
60 Queen Street
Ottawa, Ontario
Canada

William F. Hyde
Center for Environmental Policy
  Research
Duke University
212 Bioscience Building
Durham, NC 27706

H. Ishii

Emil Q. Javier
International Service for National
  Agricultural Research
P.O. Box 93375
2509 AJ, The Hague
Netherlands

Mogens Jul
Associate Professor Emeritus of the
  Royal Veterinary and Agricultural
  University
Hestehavevej 18
DK - 3400 Hillerod
Denmark

Morley R. Kare
Monell Chemical Senses Center
3500 Market Street
Philadelphia, PA 19104-3308

Robert W. Kates
Director
World Hunger Program
Brown University, Box 1831
Providence, RI 02910

Solomon H. Katz
Department of Anthropology
University of Pennsylvania
Philadelphia, PA 19104

Ambassador P. K. Kaul
Embassy of India
2107 Massachusetts Avenue NW
Washington, D.C. 20008

Eddie F. Kimbrell
U.S. Department of Agriculture
Room 3064, South Building
14th Street SW at Independence
  Avenue
Washington, D.C. 20250

Arthur Klatt
Director
CIMMYT
Londres 40, Apdo Post. 6-641
06600 Mexico, D.F.

D. C. Kothari
Kothari Buildings
114 Nungambakkam High Road
Madras 600 034
India

Mrs. D. C. Kothari
Kothari Buildings
114 Nungambakkam High Road
Madras 600 034
India

Pradip D. Kothari
Kothari General Foods Corp. Ltd.
Kothari Buildings
114 Nungambakkam High Road
Madras 600 034
India

Patricia Kutzner
Director
World Hunger Education Service
1317 G Street NW
Washington, D.C. 20005

Richard A. Ledford
Chairman
Department of Food Sciences
Cornell University
114 Stocking Hall
Ithaca, NY 14853-7201

Brian Jacques LeMay
Assistant Director
Office of International Relations
3123 S. Dillon Ripley Center
Smithsonian Institution
Washington, D.C. 20560

Martha Lewis
Director
Women in Development
Partners of the Americas
1424 K Street NW
Washington, D.C. 20005

Bjorn Lundgren
Director
International Council for Research in Agroforestry
P.O. Box 30677
Nairobi
Kenya

James MacCracken
Executive Director
Christian Children's Fund, Inc.
P.O. Box 26511
Richmond, VA 23261

Kathleen C. MacDonough
Phillip Morris Management Corporation
120 Park Avenue
New York, NY 10017

William Marion
Executive Director
CAST
137 Lynn Avenue
Ames, IA 50010-7120

G. E. Marler
Director, Corporate and Scientific Programs
IBM World Trade Americas Group
Rockwood Road
North Tarrytown, NY 10591

William B. Mashler
Assistant Secretary General
Rockefeller Foundation
4 Woody Lane
Larchmont, N.Y. 10538

Donald L. McCune
Managing Director
IFDC
P.O. Box 2040
Muscle Shoals, AL 35662

Martin McLaughlin
National Conference of Catholic Bishops
1312 Massachusetts Avenue NW
Washington, D.C. 20005-4105

Roy E. Morse
President
Institute of Food Technologies
7522 Lasater Road
Clemmons, NC 27012

Hugh T. Murphy
Vice President, Finance and Administration
Winrock International
Route 3, Petit Jean Mountain
Morrilton, AR 72110

Ignacio Narvaez Morales
Program Director
Global 2000 Inc.
Khartoum
Sudan

Marion Nestle
Chief, Nutrition Branch
Office of Disease Prevention
U.S. Department of Health and Human Services
330 C Street SW
Washington, D.C. 20201

Owen J. Newlin
Pioneer Hi-Bred International, Inc.
700 Capital Square
400 Locust Street
Des Moines, IA 50309

John J. Nicholaides III
Director, Office of International Agriculture
Associate Dean, College of Agriculture
Assistant Vice Chancellor for Research
University of Illinois at Urbana-Champaign
113 Mumford Hall
1301 West Gregory Drive
Urbana, IL 61801

Abbas Ordoobadi
International Economic Counselor
8101 Connecticut Avenue
Chevy Chase, MD 20815

Jim Phippard
Professional Staff
U.S. Senate Committee on
 Agriculture, Nutrition, and
 Forestry
SD 647
Washington, D.C. 20510

Merle Pierson
Department of Food Science and
 Technology
Virginia Tech University
Blacksburg, VA 24061

V. K. Ramachandran
W.I.D.E.R.
Annankatu 42 C
Helsinki 10
Finland

J. Thomas Ratchford
Associate Executive Director
American Association for the
 Advancement of Science
1333 H Street NW
Washington, D.C. 20005

Gordon C. Raussen
CEA Executive Office of the
 President
University of California at Berkeley
Giannini Hall
Berkeley, CA 94720

Shlomo Reutlinger
Senior Economist
Advisor on Food Security
Office of Vice President for Africa
 Region
The World Bank
1818 H Street NW, Room N 1036
Washington, D.C. 20433

Robert E. Rhoades
Leader, Food Systems
International Potato Center
P.O. Box 5969
Lima
Peru

T. H. Roberts, Jr.
Chairman
DeKalb Corp.
3100 Sycamore Road
DeKalb, IL 60115

Sunil Roy
Board Member
Center for Science and the
 Environment
890 Defence Colony
New Delhi 110 024
India

K. Russell
National 4-H Council
7100 Connecticut Avenue NW
Chevy Chase, MD 20815

Elisa Sabatini
211 Edgewood Avenue SW
Topeka, KS 66606

Setijati Sastrapradja
Director and Botanist
Centre for Research in Biotechnology
Indonesian Institute of Sciences
Bogor
Indonesia

Gunawan Satari
Director General
Agency for Agricultural Research and
 Development
Ministry of Agriculture
Jalan Ragunan 29
Jakarta 12520
Indonesia

Lowell D. Satterlee

Nevin S. Scrimshaw
International Food and Nutrition
 Program
Massachusetts Institute of
 Technology
Room E38-756
Cambridge, MA 02139

S. Bruce Shearer
President
Population Resource Center
622 Third Avenue
New York, NY 10017

Manfred Siegel
First Secretary, Agricultural Affairs
Embassy of the German Democratic
  Republic
1717 Massachusetts Avenue NW
Washington, D.C. 20036

Lisa Smith
International Rice Research Institute
P.O. Box 933
Manila
Philippines

Robert E. Smith
Senior Vice President for Research
  and Development
Nabisco Brands, Inc.
P.O. Box 1943, 100 Deforst Avenue
East Hanover, NY 07936-1943

William H. Smith
International Rice Research Institute
P.O. Box 933
Manila
Philippines

John Stoval
Chief, Research Division
BIFAD
Room 5318 New State
Agency for International
  Development
Washington, D.C. 20523

Samuel E. Stumpf
Research Professor
School of Medicine
Professor of Law
School of Law
Vanderbilt University
Nashville, TN 37240

Mrs. Samuel E. Stumpf
c/o Vanderbilt University
Nashville, TN 37240

Madhura Swaminathan
43, Alan Bullock Close
Caroline Street, St. Clements
Oxford OX 41 AU
United Kingdom

Mina Swaminathan
Delhi State Social Welfare Board
B-4/142 Safdarjang Enclave
First Floor
New Delhi 110029
India

Wayne Swegle
Director of Public Affairs and
  Communications
Winrock International
Route 3, Petit Jean Mountain
Morrilton, AR 72110

Charles Sykes
Assistant Executive Director
CARE
2025 I Street NW, Suite 1004
Washington, D.C. 20006

Mark Uebersax
Professor and Acting Chair
Department of Food Science and
  Human Nutrition
Michigan State University
East Lansing, MI 48824

Nila A. Vehar
Director
The Conference Board
1755 Massachusetts Avenue NW,
  Suite 312
Washington, D.C. 20036

Stephen Viederman
Executive Director
Jessie Smith Noyes Foundation
16 East 34th Street
New York, NY 10016

Ruben L. Villareal
Dean, College of Agriculture
University of the Philippines
Laguna 3720
Philippines

Ben J. Wallace
Professor and Chair
Department of Anthropology
Southern Methodist University
Dallas, TX 75275-0336

Qing Wang
Director
Food Research Institute
Beijing
People's Republic of China

Judy Weathers
National 4-H Council
7100 Connecticut Avenue NW
Chevy Chase, MD 20815

Tom de Wilde
Executive Director
API
1313 H Steet NW
Washington, D.C. 20005

Edward L. Williams
Coordinator, General Foods World
  Food Prize
Winrock International
Route 3, Petit Jean Mountain
Morrilton, AR 72110

Amy Wilson
International Office
American Association for
  Advancement of Science
1333 H Street NW
Washington, D.C. 20005

Yukio Yamada
Professor
Division of Tropical Agriculture
Kyoto University
Kitashirakawa Diwake-cho
Kyoto 606
Japan

Note: *Participation was by invitation. The organizers regret that some attendees who responded late to the invitation failed to have their names placed on the List of Participants.*

The title of this volume, "Science, Ethics, and Food," is the theme of a series of annual colloquia, organized at the Smithsonian Institution, in association with Winrock International and the General Foods Fund, Inc. This volume is the first in a projected series of proceedings of those colloquia.